Groundwater Governance

Groundwater Governance

Development, Degradation and Management

(A Study of Andhra Pradesh)

M. Srinivasa Reddy
Sanjit Kumar Rout
V. Ratna Reddy

RULNR

Research Unit for Livelihoods and Natural Resources (RULNR),
Centre for Economic and Social Studies (CESS), Hyderabad

RAWAT PUBLICATIONS
Jaipur • New Delhi • Bangalore • Guwahati • Kolkata

ISBN 978-81-316-0766-4

First published, 2016

Published by
Prem Rawat for **Rawat Publications**
Satyam Apts, Sector 3, Jawahar Nagar, Jaipur 302 004 (India)
Phone: 0141 265 1748/7006 Fax: 0141 265 1748
E-mail: info@rawatbooks.com
Website: www.rawatbooks.com

New Delhi Office
4858/24, Ansari Road, Daryaganj, New Delhi 110 002
Phone: 011 2326 3290

Also at *Bangalore, Guwahati* and *Kolkata*

Typeset by Rawat Computers, Jaipur
Printed at Chaman Enterprises, New Delhi

Contents

List of Tables

List of Figures and Maps

Figures

Maps

List of Photos

List of Annexures

Tables

Figures

Maps

Foreword

In a number of countries, especially in developing countries like India, groundwater is the single largest source of drinking as well as irrigation water. In the absence of scientific information with the communities, extraction of groundwater resources for productive purposes has become a risky venture leading to adverse impacts on livelihoods. Lack of information at an appropriate scale is proving to be a stumbling back for community-based groundwater management. Moreover, the externalities associated with over exploitation of groundwater resources resulting widespread failure of wells is identified as one of the main reasons for pushing farmers in to debt trap and the resultant widespread farmer suicides in India. The negative externalities are also increasingly becoming severe in the context of climate variability.

This book is an outcome of a collaborative research effort between RULNR-CESS and LNRMI, supported by Jemsetji Tata Trust (JTT) Mumbai, carried out in undivided Andhra Pradesh (AP), which is among the states where groundwater is the single largest source of irrigation as well as domestic water supplies. Besides, the state is severely affected by groundwater depletion, which is a cause of concern in terms of resource sustainability in general and groundwater and energy resources in specific. In this book an attempt has been made to understand and identify the gaps in groundwater development as well as management in terms of technical knowledge, scale, and participatory approaches.

The book is divided into two parts. While the Part I of the study attempts to assess the groundwater situation in AP using spatio-temporal analysis and highlights the importance of hydrological information at an appropriate scale to the user communities from a socio-economic perspective, Part II of the book explores the possible options for community based groundwater management in the Indian context. Based on the evidence, Part I of the book shows how groundwater is depleting along with increasing dependence over the years across the regions of AP. It is argued that the negative externalities could be mitigated to a large extent with proper dissemination of information among the communities. This must be fostered through policy support that paves the way for treating the resources as a common pool resource instead of allowing it to be exploited like a private resource.

The main focus of the Part II of the book is to understand the functioning and efficiency of groundwater management institutions by comparing and contrasting three participatory groundwater models in AP, viz. the APFAMGS, WASSAN and CWS. The study has attempted to assess the operational modalities and the impact of these institutions on access, equity and sustainability of groundwater use at the village and household level using the qualitative and quantitative information. All these three models have the common goal and objective of sustainable groundwater management, though the approaches followed and the implementation modalities are different and thus can be grouped into two broad categories, as: (i) knowledge intensive (APFAMGS); and (ii) social regulation (SRWM and APDAI). It is observed that the social regulation approach works better for sustainable groundwater management when compared to the knowledge intensive approach, as the latter is not designed to address equity. The most important lessons from these models include: (i) creation of information at appropriate scale through community involvement; and (ii) generating demand for demand management of groundwater with the help of this information.

Though these approaches have proved that communities are capable of managing groundwater in a sustainable manner, sustainability of these initiatives is a major concern in all the approaches. Besides, in the absence of contribution, the financial sustainability of the initiatives would be a big concern, especially once the external funding stops. Thus, the experience of the three models reveals that wide ranging policy changes are required to scale up the achievements of these small scale initiatives. Unless wide-ranging policy changes are brought in, these initiatives remain as models rather than being adapted at a wider scale. Replication of these models could be possible with high transaction costs, but the sustainability of these initiatives remains uncertain in the present policy environment.

Therefore, this book calls for wide-ranging policy changes so as to adapt these initiatives in a wider scale as the demand management models cannot be effective as long as policy environment is supply-sided and thus, provides valuable suggestions to policy makers. I hope it would be useful to the farmers, research community, policy makers, development practitioners and all those interested in the sustainable management of groundwater.

Professor S. Galab
Director,
Centre for Economic and Social Studies (CESS),
Nizimia Observatory Campus, Begumpet,
Hyderabad-500016, Telangana (State).

Acknowledgements

This book is an outcome of the collaborative study undertaken by Research Unit for Livelihoods and Natural Resources (RULNR), at the Centre for Economic and Social Studies (CESS), Hyderabad and Livelihoods and Natural Resources Management Institute (LNRMI), Hyderabad.

During the course of this study, we have received valuable support from various individuals and institutions, which ultimately paved the way for this book. We wish to acknowledge several of those who have provided invaluable guidance and support to us during the course of this study.

First and foremost, we humbly acknowledge our indebtedness to the Jamsetji Tata Trust, Mumbai, for providing financial support to carry out this research work. We would like to thank Prof R. Radhkrishna, Chairman of CESS and Prof. K.L. Krishna his predecessor, Prof. S. Galab, the current Director of CESS, and his predecessor Prof. Manoj Panda for all their support and encouragement during the period of this study. Our profound thanks also goes to Professor M. Gopinath Reddy, Head, Division for Sustainable Development Studies (DSDS) and the Principal Coordinator of RULNR for his continuous support, great enthusiasm and optimism, and for his deep spirit of cooperation. Thanks are also due to the members of the Research Committee, all the faculty and staff members of RULNR and CESS for their cooperation and encouragement. In particular, we would like to thank Dr Jyothis Sathyapalan, Coordinator-Dry-land eco-systems, who gave valuable guidance while carrying out this study.

We would like to thank the Department of Ground Water, Government of Andhra Pradesh and Telengana for being the partner in the 1st phase of the study. Our special thanks goes to Dr Ashok Kumar Jain, Adviser (RD), NITI Aayog, New Delhi, Mr B.M. Murali Krishna Rao, Director and Mr N. Srinivasulu, Assistant Director, Department of Groundwater, Government of Andhra Pradesh for facilitating the requisite secondary data relevant to the 1st phase of the study.

We sincerely thank Prof. R. Parthasarathy, Centre for Environmental Planning and Technology Trust (CEPT) University, Ahmadabad, Dr M. Dinesh Kumar, Executive Director,Institute for Resource Analysis and Policy, Hyderabad and Prof. A. Narayanamoorthy, Department of Economics and Rural Development, Alagappa University, Karaikudi for their constructive comments on the earlier drafts of the research. Their valuable comments have improved the quality of our work. Thanks are also due to Dr Charles Batchelor, Prof. R.S. Deshpande and Dr M.S. Rammohana Rao for useful discussions on the subject.

We would like to thank the farming communities of the sample villages who have been the main contributors to the study spending long hours in Focus Group Discussions and personnel interviews. The communities' efforts and cooperation are gratefully acknowledged. The field research was ably supported by many non-governmental organisations – CWS, WASSAN, BIRDS, DIPA and RIDS. We sincerely appreciate and thank them for their support. Our special thanks are due to Mr. Rammohan, CWS and Mr M.V. Ramachandrudu, WASSAN for their support. Thanks are also due to all the Research Investigators for their dedicated work during field study. Thanks are also extended to Prof. K.V. Reddy, Dr D. Sree Rama Raju, Mr P. Raja Narender Reddy, Mr Alex, Mr Brajraj Mishra, Dr N. Jayakumar and Mr R. Rambhupal Reddy for their support in completion of this study.

Our thanks also goes to Dr P. Prudhvikar Reddy, Prof. K.S. Reddy, Dr T.S. Jeena and Dr B. Suresh Reddy for their support and encouragement. We would like to thank all the staff members of LNRMI for their cooperation and encouragement during the course of this study. In particular, we would like to thank Dr T. Chiranjeevi for her moral support and cooperation at various stages of this research.

M. Srinivasa Reddy
Sanjit Kumar Rout
V. Ratna Reddy

Acronyms and Abbreviations Used

AARF	Actual Annual Rainfall
ACWADAM	Advanced Centre for Water Resources Development and Management
AEI	*Aide e Enfance de l Inde* (Luxembourg)
AFPRO	Action for Food Production
AP	Andhra Pradesh
APDAI	Andhra Pradesh Drought Adaption Initiative
APFMGS	Andhra Pradesh Farmer Managed Groundwater Systems
APMIP	Andhra Pradesh Micro-Irrigation Project
APSIDC	Andhra Pradesh State Irrigation Development Corporation
APTRANSCO	Andhra Pradesh Transmission Corporation Limited
APWALTA	Andhra Pradesh Water Land and Trees Act
APWELL	Andhra Pradesh Groundwater Borewell Irrigation Schemes Project
AEC	Arcadis Euro Consult
ayacut	Command area or the area served by an irrigation project such as a canal, dam or a tank
BC	Backward Caste
BCM	billion cubic metres

BIRDS	Bharathi Integrated Rural Development Society (an NGO working base at Allgadda and surrounding Mandals of Kurnool District)
BUAs	Borewell Users Associations
BPL	Below Poverty Line
BWs	Borewells
C	Command
CBGM	Community-Based Groundwater Management
CBIs	Community-Based Institutions
CBOs	Community-Based Organisations
CESS	Centre for Economic and Social Studies
CGWA	Central Ground Water Authority
CGWB	Central Ground Water Board
CI	Crop Intensity
CIGs	Common Interest Groups
Coastal Andhra	Circar Districts of Andhra Pradesh (or) is a region of India's Andhra Pradesh State
CPRs	Common Pool Resources
CRD	Commissioner of Rural Development
Cum	cubic metres
CW	Community Well
CWB	Crop Water Budgeting
CWC	Central Water Commission
CWS	Centre for World Solidarity
DDP	Desert Development Programme
de facto	actual, though not official
DIPA	Development Initiatives and Peoples Action (an NGO working base at Giddalur and surrounding Mandals of Prakasam District)
DoRD	Department of Rural Development
DPAP	Drought Prone Area Programme
DRDA	District Rural Development Agency
DT	Distribution Transformer
DWMA	District Water Management Agency

FAO	Food and Agricultural Organisation
FFS	Farmers Field Schools
FGD	Focus Group Discussions
FWS	Farmer Water Schools
GEB	Gujarat Electricity Board
GEC	Groundwater Estimation Committee
GIA	Gross Irrigated Area
GIS	Geological Information System
GMC	Groundwater Management Committee
GoAP	Government of Andhra Pradesh
GoI	Government of India
Gonchi	Collective community efforts in bringing water from a stream and distributing the same equally to irrigate a stipulated *ayacut* area
GP	*Gram Panchayat* (Village Panchayat; the third tier of the *Panchayat Raj system; Panchayat* being a representative body or Village representative body)
Grama Samakhyas	Village Organisations
GWD	Groundwater Development
Ha	Hectare or up to 2.50 acres of land
HDI	Human Development Index
HESA	Hydro-Ecosystem Analysis
HHs	Households
Hivre Bazar	A village in western Maharashtra called Hivre Bazar, where a watershed development scheme was implemented under the Adarsh Gav Yojana (AGY) of the government of Maharashtra in the mid-1990s.
HMRs	Hydrological Monitoring Records
HP	Horse Power
HPI	Human Poverty Index
HU	Hydrological Units
HUN	Hydrological Unit Network
I&CAD	Irrigation and Command Area Development

ID	Irrigable Dry
II	Irrigation Intensity
IRM&ED	Institute for Resource Management and Economic Development
IWMI	International Water Management Institute
Jal Chitra	Ability to allow a user to draw a community-based water resource map of the village
JGS	Jyoti Gram Service
Kharif	Cropping season from June to September or Agricultural Season between June to October (or) Monsoon crop growing season
Kwh	Kilowatt hour
LMF	Large and Medium Farmers (above 5 acres of land)
LNRMI	Livelihoods and Natural Resources Management Institute
Ltd.	Limited
Mahila Samkhyas	Federation of Women Thrift and Self-Help Groups
Mandal	Administrative block/unit (erstwhile block Panchayat)
MAR	Managed Aquifer Recharge
mcm	million cubic metres
MDGs	Millennium Development Goals
MF	Marginal Farmers (up to 2.5 acres of land)
mha	million hectares
MMS	*Mandal Mahila Samkhyas* (Federation of all the SHGs in the Mandal)
MoWR	Ministry of Water Resources
MP	Madhya Pradesh
NA	Not Applicable/Not Available
NABARD	National Bank for Agriculture and Rural Development
NC	Non-Command
NCIWRD	National Commission on Integrated Water Resources Development

Neerugatti	Waterman (a person dedicated to allocating water among the farmers – responsible for opening and closing the sluice and also for providing irrigation)
NGOs	Non-Governmental Organisations
NGRI	National Geographical Research Institute
NIA	Net Irrigated Area
Non-*ayacut*	Non-Command
NRM	Natural Resource Management
O&M	Operation and Maintenance
OB	Observation
OC	Other Caste
OE	Over-Exploited
OLS	Ordinary Least Squares
OW	Owned Well
Pani Panchayats	Water User Associations
PCI	Per Capita Income
PGM	Participatory Groundwater Management
PHM	Participatory Hydrological Monitoring
PRA	Participatory Rural Appraisal
PRIs	Panchayat Raj Institutions
Rabi	Cropping season from October to December or the cropping season that follows the *kharif* (November to March (or) winter crop growing season)
Rayalaseema	Ceded Districts of Andhra Pradesh in India
RIDS	Rural Integrated Development Society (an NGO working base at Garladinne and surrounding Mandals of Anantapur District)
RNE	Royal Netherlands Embassy
RULNR	Research Unit for Livelihoods and Natural Resources
S	Safe
SC	Scheduled Castes
SC	Semi-Critical
SERP	Society for Elimination of Rural Poverty

SF	Small Farmers (2.5 to 5 acres of land)
SGD	Stage of Groundwater Development
SGWD	State Ground Water Department
SPS	*Samaj Pragati Sahayog*
SPSS	Statistical Package for Social Science
sq km	square kilometre
SRWM	Social Regulations in Water Management
ST	Scheduled Tribes
Tahasildar	Revenue Official
Taluk	Administrative unit
Telangana	Formerly Nizam Territory of Andhra Pradesh, India
tmc	thousand million cubic metres
TMCs	Tank Management Committees
UN	United Nations
UP	Uttar Pradesh
UT	Union Territory
VIF	Variance Inflation Factor
VO	Village Organisation
WASSAN	Watershed Support Services and Activities Network
WS	Watershed
WSC	Watershed Committee
WSD	Watershed Development
WSDP	Watershed Development Programme
WUAs	Water Users Associations
WUG	Water User Group

PART I

Groundwater: Development, Degradation and Management

1

Introduction

Groundwater management is the most challenging part of water management. Hitherto, groundwater policies were in the lines of encouraging over-exploitation. These policies are in the nature of providing incentives for groundwater development such as subsidised credit, power, etc. While these policies helped in promoting groundwater development in the regions where groundwater development was below potential, they have led to over-exploitation of the resource in fragile resource regions. The inter-connectedness of aquifers and the linkages between surface and groundwater have far-reaching environmental impacts. If the existing models are accurate, degradation of aquifers could adversely affect stream flows and water availability of downstream water users. As a result of degradation, majority of the resource-poor farmers have lost or are losing access to water, as the water tables go down. Even when they own borewells, they cannot compete with the resource-rich farmers in deepening their wells (Reddy, 2005). That is, the poor are denied their rightful share in the Common Pool Resources (CPRs). As groundwater is the single largest source of irrigation and domestic water supply in a number of regions, its governance assumes importance and urgency.

In India the so-called water sector reforms have not looked at policies that would encourage development of institutions for resource management. Sustainable management of groundwater resources is crucial for ensuring long-term livelihood security for farmers dependent upon it. In the absence of institutions that mitigate the tendency to over-exploit CPRs, a rural agrarian system will be pushed towards extinction, especially when climate change-induced frequent droughts may create severe problems. Farming communities are perceptive about the fact that improved availability of water for irrigation significantly enhances livelihood security. Moreover, the poverty alleviation goals necessitate a focus on the specific needs of the poor, especially the women and the landless and land-poor families. The issue of how to secure the rights and entitlements to water for the poor people needs to be addressed on a priority basis. Though there are top down, state regulations on groundwater exploitation, they are inadequate and ineffective in the absence of awareness and clarity on rights and responsibilities of the user communities.

It is argued that the scientific knowledge should be shared with the farmers because of the simple reason that the sustainability of the resource lies in their hands (FAO, 2008). But the problem is with availability of accurate and reliable data. The available data is very sparse with low credibility due to the non-representativeness of the data. While farmers require data at the village (micro-watershed of about 500 hectares) level, the available data in India is based on an observation well and a rain gauge station for every 25 sq km. One of the ways towards better and sustainable groundwater management is through improving the awareness among communities and building their capacity to measure and monitor groundwater levels on a seasonal basis. However, there are no success stories of such initiatives. Most of the success stories world over pertain to tradable water rights, quotas, water prices, taxes, etc.

Objectives and Setting

This study is an attempt to examine the dynamics of groundwater development over time and across regions, and highlight the importance of reliable information at an appropriate scale.

The study also focuses on the importance of involving local communities in generating such information and managing groundwater with appropriate capacities to measure and monitor the resource. It is argued, based on the evidence, that farmers are capable of understanding and learning the technical skills of hydrological monitoring though the methodological flaws in assessment need to be addressed. The basic idea is how to demystify science and make it accessible and user friendly to the communities. Specific objectives include: (a) to examine the spatio-temporal variations in groundwater development in Andhra Pradesh (AP); (b) to discuss the relevance of the existing information to the user or farming communities; and (c) to explore the possibilities for generating reliable and useful information based on the existing experience at the ground level.

The study is based on the evidence from AP, which is among the states where groundwater is the single largest source of irrigation as well as domestic water supplies. Besides, the state is severely affected by groundwater depletion, which is a cause of concern in terms of resource sustainability in general and groundwater and energy resources in specific. Management strategies can be planned only when the resource status/potential is known and the constraints are identified. Analysis of the trends in groundwater development and the examination of categorical shifts are considered as major steps towards formulating policies and programmes that aim to increase equity and enhance the sustainability of groundwater resources. Thus, the study is expected to help frame future policy that will facilitate and promote efficient and equitable groundwater management. Uneven distribution of groundwater in AP severely affects some regions, as the drought and desert prone (DPAP and DDP) areas of AP have poor groundwater potential. These areas are characterised by large human and cattle populations which are continuously putting heavy pressure on the already fragile natural resource base for food, fodder and fuel. Any analysis which attempts to study the overall groundwater development in AP should consider these regional disparities.

The State of AP, consisting of an area of 2.75 lakh sq km, is endowed with a variety of physiographic features ranging from hills and undulating plains to a coastal deltaic environment (Map 2 in Annexure). The state has three major river basins – Godavari, Krishna and Pennar. The entire state falls under the semi-arid region of Peninsular India and is characterised by hot summers and cold winters. Geomorphologically, the state can be categorised into pedi-plains, coastal alluvial plains and hill ranges. Major constraints are imposed by the spatio-temporal variations in water availability, though in aggregate terms, the water is sufficient to meet current demands in all but the driest years (FAO, 2004). The state receives an annual rainfall of 940 mm on an average, with wide variations across the districts. It ranges from 1,200 mm in Srikakulam District to about 550 mm in Anantapur District. The majority of the rainfall (66%) is received from the south-west monsoon during June-September, while the north-east monsoon (October-December) contributes about 25 per cent of the annual rainfall. Based on the rainfall, topography and climate the state is divided into nine agro-climatic zones – high altitude and tribal zone, Krishna Zone, Godavari Zone, North Coastal Zone, Northern Telangana Zone, Central Telangana Zone, scarce rainfall zone, Southern Telangana Zone and Southern Zone (Map 3 in Annexure).

The State of AP is underlain by rock types ranging from archaean to recent alluvium with varied texture and structures. Nearly 85 per cent of the state, i.e. about 2.33 lakh sq km, is underlain by hard rocks – igneous, volcanic and metamorphic rocks, mainly granites, gneisses and khondalites in the Eastern Ghats, Cuddapah (middle upper Protozoic), Kurnool and Deccan traps (Eocene). The remaining 15 per cent of the area, i.e. 0.42 lakh sq km is underlain by soft rocks – tertiary and Gondwana sandstones and shales and alluvium of recent age. Dolerite dykes, quartz reefs, feldspathic and pegmatite veins extending from a few metres to a few kilometres cut across the country rocks at many places. The dolerite dykes have been emplaced along major prominent fractures. The vertical joints in granites are also aligned to the major direction of fractures/lineaments. Fluorite and apatite rocks contain fluoride-bearing minerals and are the main source of fluoride in groundwater. Hydro geological and geological features of AP are presented in Annexure (Map 4 & 5).

Soils play an important role in improving groundwater recharge. The soils of the state are broadly classified into red, black and alluvial (Map 6 in Annexure). Red soils cover the largest area in the state (69%) and occur widely in the Telangana and Rayalaseema regions. The black soils are in general transported by rivers. The deltaic alluvial soils, coastal alluvial soils and coastal sandy soils are formed by the riverine system. Changes in land use can have significant effects on infiltration rates through the soil surface on water retention capability of soils and on sub-surface transmissibility (Swallow et al., 2001).

The state's total renewable water resources (annual), both ground and surface water are estimated to be about 108.15 bcm (3,820 tmc), out of which about 62.29 bcm (2,200 tmc) is currently being utilised for drinking, agriculture, industry and power generation purposes. The per capita annual water resources work out to be slightly more than 1,400 cum, and the utilisation is about 800 cum (AP Water Vision, 2004). Countries or regions are considered water stressed when the annual per capita availability is between 1,000 and 2,000 cum. With the availability below 1,700 cum, a region is deemed 'water stressed'; with less than 1,000 cum, it becomes 'water scarce' and below 500 cum it becomes 'absolutely water scarce'. The current percentage of withdrawal of available water in AP is 58 per cent. As per UN indicator, if the percentage withdrawal is more than 40 per cent, the country is considered as water scarce. The average per capita water availability in AP as against India between 1951 and 2010 and the requirement for the year 2050 reflect the water stress condition in the state (Table 1.1). This calls for efforts towards efficient management supported by appropriate policy framework for proper water governance.

Table 1.1

Per Capita Availability of Water in AP Compared to India (in cum)

Year	*India*	*Andhra Pradesh*
1951	5,177	3,600
1991	2,100	1,600
2001	1,750	1,400
2010	1,588	1,400
2050	1,140	912

Source: GoI, Central Water Commission and Central Ground Water Board.
GoAP (2003), Water Conservation Mission.

About the Book

This book attempts to understand and identify the gaps in groundwater development as well as management in terms of technical knowledge, scale and participatory approaches. It is divided into two parts consisting of seven chapters. While the first part of the book (Chapters 1, 2 and 3) attempts to assess the groundwater situation in Andhra Pradesh (AP) using spatio-temporal analysis and highlights the importance of hydrological information at an appropriate scale to the user communities from a socio-economic perspective, second part of the book (Chapters 4, 5, 6 and 7) explores the possible options for community-based groundwater management using the case studies on participatory groundwater management in Andhra Pradesh.

Chapter 1 presents the introduction. Chapter 2 provides an overview of groundwater situation in AP[1] covering spatio-temporal aspects so as to identify the gaps in information available at various levels. The spatio-temporal analysis of groundwater development has been looked at from multiple dimensions and the assessments are re-emphasised from all the

1. AP is one of the few states in India that compile detailed hydrological information.

angles. The Stage of Groundwater Development (SGD) in AP over the years is analysed in this chapter on the basis of hydrological information available at the official levels in order to highlight the scale and intensity of the data generated on groundwater.

Factors influencing groundwater development are estimated in Chapter 3. However, in the absence of groundwater institutions at the district level, policy variables did not reveal any clear impact towards checking groundwater development and thus, the regression analysis does not prove to be of much help for a better understanding of groundwater management.

On this backdrop, Chapter 4 reviews the existing groundwater management practices at the policy level across the countries and also discusses the different basic management principles (viz. regulatory, economic and community-based) that are adopted for groundwater management. Though, a few water-scarce countries have reported success in the reduction of groundwater draft through regulatory measures, putting into effect such an approach and overseeing its implementation in a country of the size of India is nearly impossible. The situation is more complex in India where million of individual private tubewell owners (the number of groundwater structures in India is estimated at about 23–25 million), dispersed through the length and breadth of the country with varying groundwater availability and demand conditions, are engaged in groundwater extraction. Moreover, regulatory and economic approaches generally take an aggregate view of the situation, and often fail to capture the local-specific conditions such as geo-hydrology and socio-economic aspects of groundwater use. Hence, they are the least likely to be socially and politically viable. Similarly, these approaches do not address the issues of socio-economic equity.

As far as community-based groundwater management in India is concerned, some small-scale institutional innovations in different corners of the country are observed. However, these innovations are confined to small areas in the absence of policy support in bringing groundwater under the management regime, and the possibilities for scaling up these models have not been explored.

In this context, Chapter 5 of the book describes three institutional models that are addressing groundwater management in AP.[2] These institutions are assessed in terms of their structure, operational principles, functioning, and effectiveness in management of groundwater at the community level.

Chapter 6 describes the approach and profile of the sample villages selected from these three institutional models. The impact of the CBGM institutions at the household level is assessed in this chapter. Impact assessment is carried out at the two levels, viz. well owners and water sharing farmers across farm sizes. Impacts are assessed not only between different types of institutions but also with and without institutions, i.e. using the control village.

In Chapter 7, a comparative assessment of the strengths and weaknesses of these three institutional models that are addressing groundwater management in AP following different approaches are taken up in order to arrive at a feasible or acceptable institutional model for scaling up. All these three models have the common goal and objective of sustainable groundwater management, though the approaches followed and the implementation modalities are different and thus can be grouped into two broad categories, as: i) knowledge intensive (APFAMGS); and ii) social regulation (SRWM and APDAI). The study observes that the social regulation approach works better for sustainable groundwater management when compared to the knowledge intensive approach, as the latter is not designed to address equity. Though these three approaches have proved that communities are capable of managing groundwater in a sustainable manner, sustainability of these initiatives is a major concern in all the approaches.

---

2. The study is based on the evidence from AP, which has a long history of community groundwater management, and is one of the first states to initiate a joint well programme way back in 1987.

The experience of the three models reveals that wide ranging policy changes are required to scale up the achievements of these small scale initiatives. Unless appropriate policy changes are brought in, these initiatives remain as models rather than being adapted at a wider scale. Replication of these models could be possible with high transaction costs, but the sustainability of these initiatives remains uncertain in the present policy environment. Therefore, creating policy environment with incentive and disincentive structures in order to adapt these initiatives in a wider scale is required. For, demand management models cannot be effective as long as policy environment is supply-sided.

2

Groundwater Development in Andhra Pradesh

A Spatio-Temporal Analysis

A recent expert group report (GoI, 2007) indicated that groundwater resources in the country are under severe stress. In 2004, an alarming 28 per cent of the blocks in the country were in the category of semi-critical, critical or over-exploited, compared to only 7 per cent in 1995. Given the fact that groundwater is irrigating about 70 per cent of the cropped area and supplying 80 per cent of domestic water, it is clear that the resource is approaching a flashpoint. The rate of extraction far exceeds the rate of replenishment in many blocks, leading to a progressive lowering of the water table. The report perceives shortcomings in the legislative actions, including slowing down of development by the permit system, difficulties in enforcing regulations, scope for corruption and depriving new water users of water allocation. Unfortunately, groundwater development is still treated as a supply side issue, without any concern for demand side aspects.

Estimates of Groundwater in India

India with 2.4 per cent of the world's total area has 16 per cent of the world's population, but has only 4 per cent of the total available fresh water. This clearly indicates the need for water resources development, conservation and their optimum use. At the aggregate level, India is not short of water. The water resources potential of the country has been assessed from time to time by different agencies (Table 2.1). It may be seen that since 1954, the estimates have stabilised and are within the proximity of the currently accepted estimate of 1,869 billion cubic metres (bcm), which includes replenishable groundwater that gets charged on annual basis.

Table 2.1

Estimates of Water Resources of India

Agency	*Estimate (in bcm)*	*% Deviation (from 1,869 bcm)*
First Irrigation Commission (1902–03)	1,443	–23
Dr A.N. Khosla (1949)	1,673	–10
Central Water and Power Commission (1954–66)	1,881	+ 0.6
National Commission on Agriculture	1,850	– 1
Central Water Commission (1988)	1,880	+ 0.6
Central Water Commission (1993)	1,869	–

Source: GoI (2007), Report of the Steering Committee on Water Resources for Eleventh Five Year Plan (2007–2012), Planning Commission, May.

Within the limitations of physiographic conditions, socio-political environment, legal and constitutional constraints and the technology available at hand, the utilisable water resources of the country have been assessed at 1,123 bcm, of which 690 bcm is from surface water and 433 bcm from groundwater sources (CWC, 1993). Harnessing of 690 bcm of utilisable surface water is possible only if matching storages are built. Trans-basin transfer of water, if taken up to the full extent as proposed under the National Perspective Plan, would further increase the utilisable quantity by approximately 220 bcm. The irrigation potential of the country has been estimated at 139.9 million hectares (mha) without inter-basin sharing of water and 175 mha with inter-basin sharing. The requirement of water for various sectors has been assessed by the National Commission on Integrated Water Resources Development (NCIWRD) in the year 2000. This requirement is based on the assumption that the irrigation efficiency will increase to 60 per cent from the present level of 35 per cent to 40 per cent. The Standing Committee of the Ministry of Water Resources (MoWR) also assesses it periodically (Table 2.2).

Table 2.2

Water Requirement for Various Sectors

Sector	*Water Demand in km³ (or bcm)*					
	Standing Sub-Committee of MoWR			*NCIWRD*		
	2010	*2025*	*2050*	*2010*	*2025*	*2050*
Irrigation	688	910	1,072	557	611	807
Drinking Water	56	73	102	43	62	111
Industry	12	23	63	37	67	81
Energy	5	15	130	19	33	70
Others	52	72	80	54	70	111
Total	813	1,093	1,447	710	843	1,180

Source: GoI (2007), Report of the Steering Committee on Water Resources for Eleventh Five Year Plan (2007–2012), Planning Commission, May.

The annual replenishable groundwater resource for the entire country is 431.03 bcm (Table 2.3). The overall contribution of rainfall to the country's annual replenishable groundwater resource is 67 per cent and the share of other sources, including canal seepage, return flow from irrigation, seepage from water bodies and water conservation structures taken together is 33 per cent. In the states of AP, Delhi, Haryana, Jammu & Kashmir, Jharkhand, Punjab, Tamil Nadu, Uttar Pradesh, Uttarakhand and the UT of Puducherry, the contribution of other sources is more than the national average of 33 per cent, mainly because of canal seepage and intensive irrigation. The southwest monsoon being the most prevalent contributor of rainfall in the country, about 73 per cent of the country's annual replenishable groundwater recharge takes

Map 2.1

Categorisation of Blocks/Mandals/Taluks in India (as on 31st March 2009)

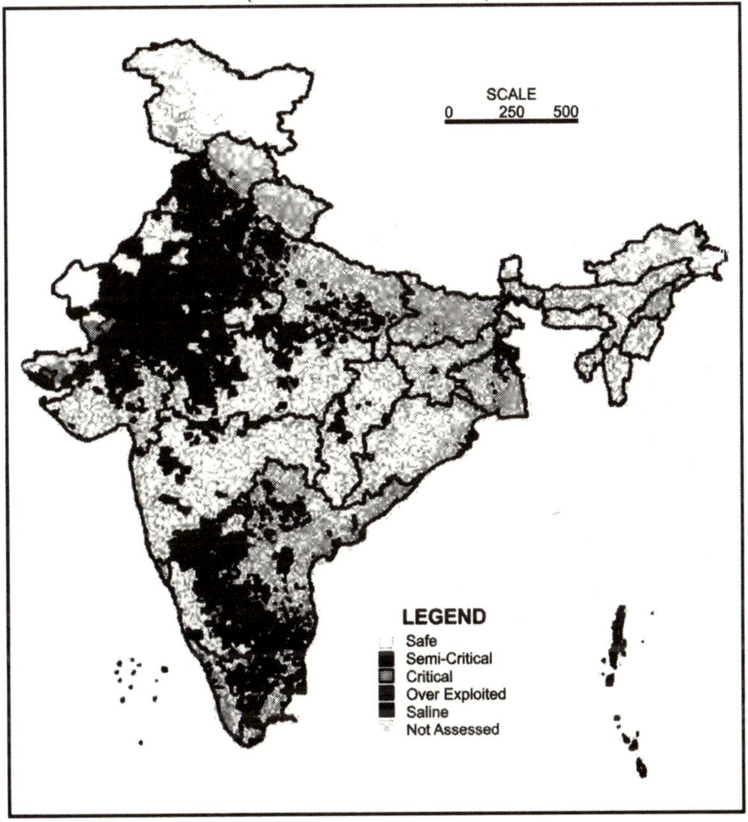

Source: GoI (2011), Dynamic Groundwater Resources of India (as on 31 March 2009), Central Ground Water Board, Ministry of Water Resources, Faridabad, November.

place during the *kharif* period of cultivation. Keeping 35.03 bcm as the allocation for natural discharge during the non-monsoon season, the net annual groundwater available for utilisation in the entire country is about 396.06 bcm. The state-wise groundwater resources availability, utilisation and categorisation of over-exploited and critical blocks are given in Table 2.3 to assess the relative status of AP.

Table 2.3

State-wise Groundwater Resources Availability, Utilisation and Status of Groundwater Development in India

States/Union Territories	Annual Replenishable Ground Water Resource	Natural Discharge during Non-Monsoon Season	Net Annual Ground Water Availability	Annual Ground Water Draft	Stage of Ground Water Development (%)	Categorisation of Assessment Units (in Number)	
						Over-Exploited	Critical
Andhra Pradesh	33.83	3.07	30.76	14.15	46	219	77
Arunachal Pradesh	4.45	0.45	4.01	0.003	0.07	0	0
Assam	30.35	2.537	27.81	6.026	22	0	0
Bihar	28.63	2.42	26.21	11.36	43	0	0
Chhattisgarh	12.22	0.64	11.58	3.6	31	0	0
Delhi	0.31	0.02	0.29	0.4	138	7	0
Goa	0.221	0.088	0.133	0.044	33	0	0
Gujarat	18.43	1.08	17.35	12.99	75	31	12
Haryana	10.48	0.68	9.8	12.43	127	55	11
Himachal Pradesh	0.59	0.06	0.53	0.31	58	0	0

Cont'd...

...Cont'd

Jammu & Kashmir	3.7	0.37	3.33	0.73	22	0	0
Jharkhand	5.96	0.55	5.41	1.61	30	0	0
Karnataka	16.81	2	14.81	10.01	68	65	3
Kerala	6.62	0.59	6.03	2.81	47	5	15
Madhya Pradesh	33.95	1.7	32.25	17.99	56	24	5
Maharashtra	35.73	1.93	33.81	16.95	50	7	1
Manipur	0.44	0.04	0.4	0.004	1	0	0
Meghalaya	1.2343	0.1234	1.1109	0.0017	0.15	0	0
Mizoram	0.044	0.004	0.039	0.0004	1	0	0
Nagaland	0.42	0.04	0.38	0.008	2.14	0	0
Orissa	17.78	1.09	16.69	4.36	26	0	0
Punjab	22.56	2.21	20.35	34.66	170	103	5
Rajasthan	11.86	1.07	10.79	14.52	135	140	50
Sikkim	–	–	0.046	0.01	21	0	0
Tamil Nadu	22.94	2.29	20.65	16.56	80	142	33
Tripura	2.97	0.23	2.74	0.16	6	0	0
Uttar Pradesh	75.25	6.68	68.57	49.48	72	37	13

Cont'd...

...Cont'd

Uttarakhand	2.17	0.1	2.07	1.05	51	2	0
West Bengal	30.5	2.92	27.58	10.91	40	0	1
Total (States)	**430.45**	**34.99**	**395.52**	**243.14**	**61**	**837**	**226**
Andaman & Nicobar	0.31	0.012	0.298	0.011	4	0	0
Chandigarh	0.022	0.002	0.02	0	0	0	0
Dadra & Nagar Haveli	0.059	0.003	0.056	0.009	15	0	0
Daman & Diu	0.012	0.001	0.011	0.011	99	1	0
Lakshadweep	0.0105	0.007	0.0035	0.0026	74	0	0
Puducherry	0.171	0.017	0.154	0.15	98	1	0
Total (Union Territories)	**0.59**	**0.04**	**0.54**	**0.18**	**34**	**2**	**0**
Grand Total	**431.03**	**35.03**	**396.06**	**243.32**	**61**	**839**	**226**

Source: GoI (2011), Dynamic Groundwater Resources of India (as on 31 March 2009), Central Ground Water Board, Faridabad, November.

Status of Groundwater in AP

In this sub-section, we examine the hydrological information that is available at the official level in AP. The official data sources are used to highlight the scale and intensity of the data generated on groundwater with the objective of identifying the gaps in information available at various levels. AP is one of the few states in India that compile detailed hydrological information. The source-wise composition of irrigation varies across regions. While the rain-fed and drought-prone regions have experienced a shift towards groundwater irrigation, the river basin endowed regions have continued to depend on surface water resources.

In AP, the first groundwater resource estimation was undertaken in the year 1985 and subsequently the same exercise was carried out six times till 2009. The Stage of Groundwater Development (SGD) in AP over the years is analysed on the basis of the groundwater resource estimation made by the State Ground Water Department (SGWD)*. These estimates are available at the district and regional level on the basis of command and non-command areas and also on the basis of drought prone (DPAP) and non-drought prone areas. The estimates are based on the readings from the Observation (OB) Wells or assessment units located at the *Taluk*/*Mandal*/Groundwater Basin/Watershed level (Table 2.4). The assessment unit is not same over the years. The number of assessment units was only 47 *Taluks* in 1985, though the coverage was expanded to all the *Mandals*, Groundwater Basins and Watersheds since 1993. These units are assumed to be valid for 26,586 villages spread over a geographical area of 0.28 million sq km in the state. Given the high spatial variations in the structure and quality of geo-hydrology and aquifers, the relevance of district or *Mandal* level data is quite dubious.

---

* See Maps 7–17 in Annexure for estimated year-wise SGD in AP.

Table 2.4

Details of Groundwater Assessment in Andhra Pradesh

Year of Assessment	*Assessment Unit*	*No. of Units Assessed*	*Methodology*	*Actual Number*
1985	Taluk	47	Water table fluctuation	308
1993	Mandal	1,108	Water table fluctuation and rainfall infiltration	1,124
2002	Groundwater Basin	1,157	GEC 1997 methodology	1,157
2004	Watershed	1,229	GEC 1997 methodology	1,229
2007	Watershed	1,229	GEC 1997 methodology	1,229
2009	Watershed	1,229	GEC 1997 methodology	1,229

Source: GoAP, Groundwater Resource Estimated Reports (different years), Groundwater Department, Hyderabad.

The methodology adopted in Groundwater Estimation Committee (GEC) 1997 is reasonably valid in an approximately homogenous hydrologic terrain like alluvium. However, this may not be applicable for hard rock terrain where the hydro-geological conditions vary widely within small areas under the prevailing heterogeneous set up. Methodological refinements may be needed in this regard. Significantly, almost two-thirds of the area, including AP, is occupied by hard rock terrain (GoI, 2002). A committee appointed in 2001 has suggested modifications to the GEC–1997 methodology. The committee also left some important issues, such as norms of recharge components, return flow from irrigation, groundwater draft, base flow, spring discharge and specific yield, unresolved, and recommended for further studies and estimates. As a result, the methodology used for estimation is neither perfect nor appropriate for addressing the needs of the users, who need to know the actual groundwater available in their village on a season to season basis.

Groundwater resource of the state is estimated on a regular basis by the MoWR in close collaboration with the Ground Water Department of Government of AP. The administrative set up of the state was reconstituted into *Mandals* in 1985. Accordingly, groundwater resources in the state were estimated in 1995, following the norms recommended by the GEC 1984, taking 1993 as a base year. In 1997, a detailed methodology, along with a guide book giving all the computations needed for assessment was published by the GEC, a high power committee of the MoWR. This is often referred as the GEC 1997 methodology. In 2004, groundwater resource estimation using data of 2001 was completed, based on GEC 1997. Based on the recommendations of the 'Groundwater Estimation Committee on Hard Rock Terrain', resource estimation was carried out again in 2004, 2007 and 2009. As per the methodology followed, the status of groundwater is given simply as a ratio of the utilisation and recharge, which is called the SGD. It can also be called the stage of groundwater utilisation for clarity.

The groundwater status was estimated for the year 1984–85 using the water table fluctuation method. The data of Central Ground Water Board (CGWB) observation well network, supplemented by SGWD observation well data, were used. All the calculations were made for the year 1984–85. The number of observation wells monitored in the year 1984–85 by CGWB and SGWD are 321 and 2,698 respectively. The district-wise groundwater development in the state ranges between 6 per cent and 59 per cent and that of entire state is of the order of 28 per cent. Recharge computations have been made separately for *ayacut* (command) and non-*ayacut* (non-command) areas. It is observed that most of the groundwater development is confined to the non-*ayacut* areas.

The total dynamic groundwater resources of AP were thus estimated at 25.3 bcm per annum as in 1984–85 and the net annual groundwater draft was 7.07 bcm (Table 2.5). Thus, a balance of 18.23 bcm was available for future development. It is to be remembered that these estimates consider only the dynamic groundwater resources of water table aquifers.

Table 2.5

Estimates of Groundwater in Different Assessment Years (in bcm)

Year	*Annual Availability*	*Annual Utilisation*	*Balance*	*SGD*
1985	25.30	7.07	18.23	28
1993	35.29	10.13	25.16	29
2002	30.56	12.97	17.59	43
2004	32.76	14.86	17.90	45
2007	34.70	14.11	20.59	41
2009	30.76	14.15	16.61	46

Note: Data compiled from Groundwater Resource Estimated Reports of different years, Ground Water Department, GoAP.

The net groundwater availability per annum, as per 1993 estimates, for the entire state was estimated to be about 35.3 bcm, which was 14.4 per cent of the total quantity of water received through normal precipitation. From this, about 15 per cent, i.e. 5.3 bcm was earmarked for drinking and other committed uses, leaving a balance of 30 bcm for irrigation. The net annual groundwater draft for irrigation was 7.09 bcm. The level of groundwater development across districts ranged between 7 per cent and 41 per cent, and for the state as a whole it was 29 per cent (Table 2.7). However, during this period, 5 *Mandals* were categorised under dark/critical and 60 *Mandals* under semi-critical zones.

In 2002 the state was divided into 1,157 assessment units, which include basins with defined hydrological boundaries in hard rock areas with areas ranging between 50 and 450 sq km and *Mandals* (administrative blocks) in alluvial areas including 36 Saline *Mandals*. Computations of net groundwater availability, its utilisation and availability for future use in all the assessment units for command, non-command and poor groundwater quality areas were made separately. The estimates showed the groundwater availability at 31 bcm, usage at 13 bcm and the balance at 18 bcm per annum. This clearly indicates that these aggregate estimates do not reflect micro-realities where declining water levels and drying up wells is observed.

The watershed boundaries were revised to 1,229 during 2004. The estimates showed that groundwater availability was 32.8 bcm, usage was 14.9 bcm and the balance was 17.9 bcm per annum. These estimates included 1.3 bcm of net annual groundwater availability in poor quality and saline areas. The usage in saline areas was about 0.21 bcm. In comparison with 2002 estimates, there was a definite increase (by about 13%) in groundwater usage across sectors. This was corroborated by a steep decline in the mean water levels almost everywhere in the state. In many areas, water level stands in fractured formation, rather than in weathered formation, as shown by the network of existing Piezometers, and the drying up of traditional OB Wells. Groundwater development was at the highest level (45%) during 2004 due to the prevailing unprecedented drought conditions. The situation eased by 2007 with consecutive good monsoons resulting in a decline in the SGD (41%). But the situation reverted in 2009 with SGD reaching 46 per cent. This indicates a secular growth in groundwater development over the years except when there are good rainfalls in consecutive years.

Groundwater Development in Command and Non-command Areas

In 2009, estimates were made separately for command and non-command areas using the GEC 1997 methodology, based on the data from Transmission Corporation of Andhra Pradesh Limited (APTRANSCO), Revenue Department and Irrigation Department. Correction factors were applied based on the field observations. The total groundwater resources were estimated at 30.76 bcm (18.41 in non-command area and 12.35 in command area) and utilisation was 14.15 bcm (10.44 in

Table 2.6

Assessment Year-wise No. of Basins/Watersheds/Mandals and Villages Under Different Categories

Category	*Number of Basins/Watersheds*				*Number of Mandals*				*Number of Villages*			
	2002	*2004*	*2007*	*2009*	*2002*	*2004*	*2007*	*2009*	*2002*	*2004*	*2007*	*2009*
Over-Exploited	118	215	132	94	185	211	111	84	1,481	4,190	5,096	2,123
Critical	–	85	89	49	–	71	60	26	–	–	1,064	–
Semi-Critical	–	208	175	102	–	167	160	93	–	–	2,632	–
Safe	–	721	833	984	–	658	777	905	–	–	17,219	–

Source: Data compiled from GoAP, Groundwater Resource Estimated Reports of different years, Ground Water Department.

non-command area and 3.71 in command area), while the balance available resource was 16.61 bcm (7.98 in non-command area and 8.64 in command area). The average SGD for the entire state was 46 per cent, of which 57 per cent was in non-command areas while 30 per cent was in command areas (Table 2.7).

The state has been categorised into four zones, viz. safe (<70%), semi-critical (70% to 90%), critical (90 to 100%) and over-exploited (>100%), based on the percentage of groundwater exploitation. About 5,096 villages, spread over 111 *Mandals* and 132 watersheds, fell in the over-exploited category in 2007 (Table 2.6). The number of over-exploited villages had come down to 2,123 by 2009. Along with the overall groundwater development at the state and district level (Figure 2.1), the variation between command and non-command areas was also examined. The assessment shows that groundwater resources have reached a very critical stage in non-command areas (Table 2.7). All the areas of the state that are not served by canal command, including the areas in districts like West Godavari, Anantapur, etc. are showing very high usage of the available groundwater and this is reflected in the SGD, which exceeds 70 per cent of the safe limit of exploitation. The total groundwater resources of the state during 2009 were estimated at 18.41 bcm in non-command areas and 12.34 bcm in command areas, while the utilisation was 10.43 bcm in non-command areas and 3.70 bcm in command areas, and the balance available resource was 7.98 bcm in non-command areas and 8.64 bcm in command areas (Table 2.8).

The estimates show that the overall SGD in AP has gone up from 28 per cent in 1985 to 46 per cent in 2009, except during the years 1993 and 2007, when it declined marginally. However, opposite trends were observed in the case of command and non-command areas between 1985 and 2002 while the command areas had experienced a declining trend, the non-command areas had experienced an increasing trend. However, both command and non-command areas had shown an increasing trend between

Table 2.7

District/Region-wise Stage/Level of Groundwater Development in AP

District/Region	*Stage/Level of Groundwater Development (%)*																	
	1985			*1993*			*2002*			*2004*			*2007*			*2009*		
	C	*NC*	*T*	*C*	*NC*	*T*	*C*	*NC*	*T*	*C*	*NC*	*T*	*C*	*NC*	*T*	*C*	*NC*	*T*
Srikakulam	1	37	10	18	17.9	18	15	23	21	10	10	10	19	35	28	20	37	29
Vizianagaram	2	20	9	13.1	13.1	13.1	15	30	25	19	28	24	16	24	21	15	24	21
Visakhapatnam	3	24	15	15.8	15.8	15.8	12	32	27	71	29	34	56	19	23	59	28	32
East Godavari	12	33	18	13.2	13.2	13.2	15	40	22	14	71	31	14	36	20	25	13	22
West Godavari	19	61	35	23.7	23.7	23.7	5	115	74	3	94	48	6	75	28	12	61	36
Krishna	9	56	23	15.9	15.9	15.9	18	35	26	14	53	24	19	70	29	24	73	33
Guntur	6	12	8	6.5	6.6	6.5	6	30	9	9	57	10	10	43	11	17	45	19
Prakasam	3	19	13	13.6	13.6	13.6	6	54	26	32	76	57	14	64	34	15	48	29
Nellore	13	54	32	32.5	32.5	32.5	33	49	41	36	53	44	29	53	38	25	50	37
Coastal Andhra	**8**	**34**	**18**	**16.5**	**19.6**	**17.8**	**13**	**49**	**28**	**16**	**53**	**29**	**15**	**47**	**25**	**21**	**43**	**29**
Kurnool	3	18	12	17.6	17.5	17.5	13	36	27	21	59	42	30	48	37	33	35	34
Anantapur	10	41	32	36.9	35.9	36.9	35	80	73	30	97	76	58	103	90	42	87	78

Cont'd...

...Cont'd

Cuddapah	15	58	46	35	35	35	13	64	56	43	82	74	27	87	69	40	67	62
Chittoor	8	88	44	33.7	33.7	33.7	NA	94	94	NA	72	72	NA	67	67	NA	75	75
Rayalaseema	**8**	**51**	**34**	**29.1**	**31.6**	**31.1**	**17**	**72**	**63**	**28**	**78**	**66**	**36**	**76**	**65**	**36**	**71**	**64**
Ranga Reddy	32	52	50	36.4	36.4	36.4	20	60	58	NA	103	103	NA	99	99	NA	74	74
Nizamabad	5	90	33	35.2	35.2	35.2	NA	81	81	71	96	86	69	79	75	69	81	76
Medak	10	62	43	30.8	30.8	30.8	NA	71	71	NA	87	87	NA	74	74	NA	84	84
Mahbubnagar	15	51	41	33.2	33.3	33.3	52	94	83	34	73	62	15	61	49	17	56	48
Nalgonda	5	55	31	27.2	27.2	27.2	32	27	28	11	77	52	18	73	53	24	76	56
Warangal	16	71	55	33.9	33.9	33.9	19	67	41	56	80	69	59	70	65	85	68	74
Khammam	5	14	12	10	10	10	25	53	46	22	19	20	25	16	18	36	28	30
Karimnagar	11	104	59	40.8	40.8	40.8	15	20	18	34	79	53	17	81	47	42	53	48
Adilabad	3	7	6	13.1	13.2	13.2	10	76	57	33	33	33	28	32	31	43	25	27
Telangana	**9**	**52**	**36**	**29.1**	**27.8**	**28.3**	**22**	**58**	**49**	**36**	**66**	**56**	**33**	**58**	**51**	**47**	**57**	**55**
Andhra Pradesh	**8**	**46**	**28**	**21.1**	**25.6**	**23.6**	**16**	**58**	**42**	**22**	**65**	**45**	**21**	**59**	**41**	**30**	**57**	**46**

Source: 1. Data compiled from Groundwater Resource Estimated Reports of different years, Ground Water Department, GoAP.

Note: C = Command; NC= Non-Command; T=Total (both command and non-command taken together).

2002 and 2004, which was due to the consecutive droughts during that period. Between 2004 and 2009, there was a decline in the level of development, reverting back to 2002 levels especially in the case of non-command areas. On the other hand, there was an increase in the command areas.

The overall SGD in Coastal Andhra was lesser when compared to the other two regions in all the groundwater resource estimated years. On the other hand, in Rayalaseema region the SGD was higher compared to the other two regions during these years, except in 1985, when Telangana was marginally higher than this region. The trend with respect to the SGD in command areas was same in all the three regions during the estimated years 1985 to 2002, when it increased from 1985 to 1993 and then declined. After 2002, the trend was same in case of Coastal Andhra and Telangana Regions (increased between 2002 and 2004 and declined between 2004 and 2009) while the SGD in Rayalaseema showed an increasing trend during both the periods. The non-command areas of all the three regions exhibited a similar trend with respect to the overall SGD.

On the basis of overall SGD, districts were categorised as 'very high usage' (>70%), 'high usage' (>50% & <70%), 'moderate usage' (>30% & <50%) and 'low usage' (<30%) districts. As per Groundwater resource estimation of 1985, none of the district from any region came under the 'very high usage' category. Three of the districts in the Telangana Region (Ranga Reddy, Warangal and Karimnagar) came under the 'high usage' category; while two districts from Coastal Andhra (West Godavari and Nellore), three from Rayalaseema (Anantapur, Cuddapah and Chittoor) and four from Telangana (Nizamabad, Medak, Mahbubnagar and Nalgonda) were under the 'moderate usage' category; and seven Coastal Andhra districts (Srikakulam, Vizianagaram, Visakhapatnam, East Godavari, Krishna, Guntur and Prakasam), one Rayalaseema district (Kurnool) and two Telangana districts (Khammam and Adilabad) were under the 'low usage' category.

Figure 2.1

District/Region-wise Level of Groundwater Development in AP(%)

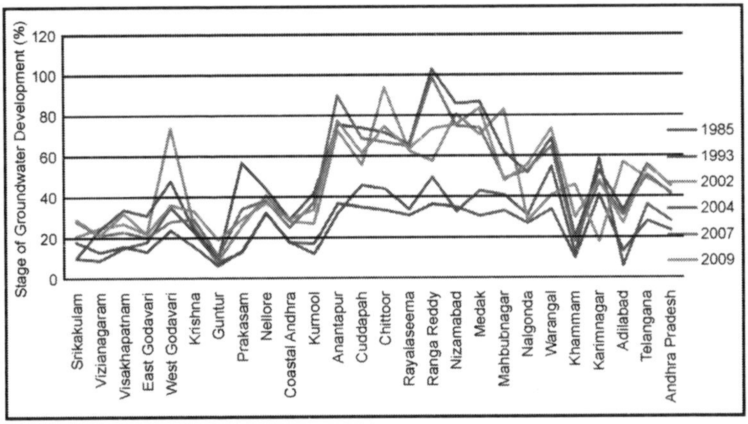

Groundwater development estimates in 1993 showed that none of the districts from any region came under the 'very high' and 'high usage' category. Only one district from the Coastal Andhra (Nellore), three in Rayalaseema (Chittoor, Cuddapah and Anantapur) and six districts from the Telangana Region (Mahbubnagar, Ranga Reddy, Medak, Nizamabad, Karimnagar and Warangal) came under the 'moderate usage' category; while all the districts from Coastal Andhra, only one from Rayalaseema (Kurnool) and three from the Telangana Region (Adilabad, Khammam and Nalgonda) came under the 'low usage' category. During 2002, one district in Coastal Andhra (East Godavari), two in Rayalaseema (Chittoor and Cuddapah) and three in the Telangana Region (Ranga Reddy, Medak and Nizamabad) were found to be under the 'very high usage' category. The 'high usage' districts included one in Rayalaseema (Anantapur) and two in Telangana (Mahbubnagar and Nalgonda). The 'moderate usage' category included one district in Coastal Andhra (Nellore) and two in Telangana (Karimnagar and Warangal), while seven districts in Coastal Andhra (Srikakulam, Vizianagaram, Visakhapatnam, East Godavari, Krishna, Guntur and Prakasam), only one in

Rayalaseema (Kurnool) and two in Telangana (Adilabad and Khammam) were under the 'safe/low usage' category.

The severe drought conditions preceding 2004 reflected in three districts in Telangana (Ranga Reddy/Hyderabad, Nizamabad and Medak) and three in Rayalaseema (Anantapur, Cuddapah and Chittoor), falling under the 'very high usage' category in 2004. One district in Coastal Andhra (Prakasam) and four in Telangana (Warangal, Mahbubnagar, Karimnagar and Nalgonda) came under the 'high usage category', while four districts in Coastal Andhra (Visakhapatnam, East Godavari, West Godavari and Nellore), one in Rayalaseema (Kurnool) and one in Telangana (Adilabad) were under the 'moderate usage' category. The 'safe/low usage' districts were Vizianagaram, Srikakulam, Krishna and Guntur from Coastal Andhra (four districts), and only one from Telangana (Khammam). The situation continued during 2007, when the number of 'very high usage' districts comprised three from Telangana (Ranga Reddy/Hyderabad, Nizamabad and Medak) and one from Rayalaseema (Anantapur). The 'high usage' districts included two districts from Rayalaseema (Cuddapah and Chittoor) and two from the Telangana Region (Warangal and Nalgonda). Two districts in Coastal Andhra (Nellore and Prakasam), one in Rayalaseema (Kurnool) and three in Telangana (Mahbubnagar, Karimnagar, Adilabad) came under the 'moderate usage' category. As many as seven districts from Coastal Andhra (Krishna, Srikakulam, West Godavari, Visakhapatnam, Vizianagaram, East Godavari and Guntur) and only one in Telangana (Khammam) were under the 'low/safe usage' category.

Groundwater development estimates in 2009 showed that none of the districts from Coastal Andhra region came under the 'very high' and 'high usage' category. On the other hand, none of the districts from Rayalaseema was found to be in 'safe/low usage' category. Two districts in Rayalaseema (Anantapur and Chittoor) and four in Telangana Region (Ranga Reddy, Nizamabad , Medak and Warangal) were found to be under the 'very high usage' category. Two districts, one in Rayalaseema

(Kadapa) and one in Telangana (Nalgonda) fall under the 'high usage' category. The 'moderate usage' category included four districts in Coastal Andhra (Visakhapatnam, West Godavari, Krishna and Nellore), one in Rayalaseema (Kurnool) and two in Telangana (Mahbubnagar and Karimnagar), while five districts in Coastal Andhra (Srikakulam, Vizianagaram, East Godavari, Guntur and Prakasam), two in Telangana (Adilabad and Khammam) were under the 'safe/low usage' category (Table 2.7).

Groundwater Development in DPAP and Non-DPAP Districts

When the districts are categorised as DPAP and non-DPAP districts (see annexure Map 18), the trends in groundwater development shows that the overall SGD in DPAP districts are higher than that of non-DPAP districts in all the assessment years except in 1985, when non-DPAP districts had a marginally higher level of groundwater development (Table 2.8). While command areas of DPAP districts had higher level of groundwater development than the non-DPAP districts during 1993, 2004 and 2007, in other assessment years (1985, 2002 and 2009) the reverse was found (command areas of non-DPAP districts had higher level of groundwater development than that of DPAP districts). The command areas of DPAP and non-DPAP districts exhibited a similar trend in all the years, i.e. increase in 1993, 2004 and 2009 and decline in 2002 and 2007, when compared to the respective previous years' estimates. As far as non-command areas of DPAP and non-DPAP districts are concerned, except in 2007 and 2009, non-DPAP districts had higher level of groundwater development than the DPAP districts. Both command and non-command areas within the respective zones showed similar trends. The severe drought conditions during 2004 had led to groundwater exploitation (increased utilisation), especially in the non-DPAP districts and hence 2004 showed higher groundwater balance despite increase in use.

Overall, the SGD of most districts was under the safe category in 1985 and 1993 (see Annexure Table 1 and Figure 1 & 2). In

Table 2.8

Groundwater Estimates for DPAP and Non-DPAP Districts in AP

Scheme/ Year	Annual Groundwater Availability (mcm)			Groundwater Utilisation/Draft (mcm)			Groundwater Balance (mcm)			SGD (%)		
	C	NC	Total	C	NC	Total	C	NC	Total	C	NC	Total
DPAP												
1985	4,832	7,774	12,606	298	3,047	3,345	4,534	4,727	9,261	6	39	27
1993	4,562	9,747	14,307	1,019	2,461	3,481	3,542	7,286	10,823	22	25	24
2002	3,423	12,369	15,792	437	7,122	7,559	2,986	5,247	8,233	13	58	48
2004	3,918	11,473	15,391	963	7,302	8,265	2,955	4,171	7,126	25	64	54
2007	4,760	11,405	16,165	1,153	6,795	7,948	3,607	4,611	8,218	24	60	49
2009	4,026	12,253	16,279	1,077	7,165	8,242	2,949	5,088	8,037	27	58	51
Non-DPAP												
1985	7,261	5,435	12,696	703	3,025	3,728	6,558	2,410	8,968	10	56	29
1993	8,415	7,277	15,690	1,722	1,895	3,612	6,693	5,382	12,078	20	26	23
2002	7,825	6,943	14,768	1,315	4,101	5,416	6,510	2,842	9,352	17	59	37
2004	11,048	6,319	17,367	2,366	4,223	6,589	8,682	2,096	10,778	21	67	38

Cont'd...

...Cont'd

2007	12,054	6,481	18,535	2,433	3,731	6,164	9,621	2,749	12,370	20	58	33
2009	8,319	6,161	14,481	2,629	3,273	5,902	5,690	2,888	8,578	32	53	41
AP												
1985	12,093	13,209	25,303	1,001	6,073	7,074	11,092	7,737	18,229	8	46	28
1993	12,975	17,024	29,997	2,740	4,356	7,093	10,235	12,668	22,904	21	26	24
2002	11,248	19,312	30,560	1,752	11,223	12,975	9,496	8,089	17,585	16	58	42
2004	14,966	17,792	32,758	3,329	11,525	14,854	11,637	6,267	17,904	22	65	45
2007	16,814	17,886	34,700	3,586	10,526	14,112	13,229	7,360	20,588	21	59	41
2009	12,345	18,414	30,760	3,706	10,438	14,145	8,639	7,976	16,615	30	57	46

Source: 1. Data compiled from Groundwater Resource Estimated Reports different years, Ground Water Department, GoAP.

Note: C=Command (*ayacut*); NC=Non-Command (non-*ayacut*); Total=Overall (both command and non-command areas together).

Table 2.9

Region-wise Percentage of Assessment Units (Mandals) Falling Under Different Categories in AP

Year/ Region	Safe			Semi-Critical			Critical			Over-Exploited		
	C	NC	Overall	C	NC	Overall	C	NC	Overall	C	NC	Overall
1985												
Coastal Andhra	NA	NA	92.0	NA	NA	3.0	NA	NA	2.8	NA	NA	2.6
Rayalaseema	NA	NA	79.0	NA	NA	7.3	NA	NA	8.5	NA	NA	5.6
Telangana	NA	NA	73.0	NA	NA	21.0	NA	NA	2.9	NA	NA	2.7
Andhra Pradesh	NA	NA	81.0	NA	NA	11.2	NA	NA	4.1	NA	NA	3.2
1993												
Coastal Andhra	NA	NA	92.0	NA	NA	1.6	NA	NA	0.5	NA	NA	0.2
Rayalaseema	NA	NA	90.0	NA	NA	6.0	NA	NA	3.0	NA	NA	0.0
Telangana	NA	NA	90.0	NA	NA	7.6	NA	NA	1.3	NA	NA	0.4
Andhra Pradesh	NA	NA	91.0	NA	NA	5.0	NA	NA	1.4	NA	NA	0.3
2002												
Coastal Andhra	100	100	92.0	1.1	8.7	2.6	0.0	2.1	1.2	0	6.6	3.7
Rayalaseema	100	52.6	60.7	0.0	16.8	12.0	1.6	10.8	9.4	0	20.7	17.9
Telangana	85.7	61.9	68.5	1.5	21.2	15.7	0.0	5.8	4.7	0	12.8	10.7

Cont'd...

...Cont'd

Andhra Pradesh	99.4	72.9	75.9	1.0	16.3	9.8	0.2	5.9	4.3	0	12.8	9.6
2004												
Coastal Andhra	92.9	75.3	86.4	3.9	9.4	5.9	0.0	3.1	1.6	3.2	11.5	6.1
Rayalaseema	96.7	38.4	47.0	0.0	20.3	20.5	0.0	7.8	6.0	3.3	33.6	26.5
Telangana	89.5	44.9	54.8	5.3	21.6	20.1	1.5	10.2	8.1	3.8	22.8	16.8
Andhra Pradesh	92.5	52.5	65.3	3.8	17.6	14.7	0.4	7.5	5.1	3.4	22.0	14.7
2007												
Coastal Andhra	97.6	80.6	90.2	0.7	11.5	5.6	0.0	3.5	1.2	1.0	3.8	2.3
Rayalaseema	84.8	35.8	45.7	1.5	23.9	18.8	3.0	11.1	10.7	10.6	29.2	24.8
Telangana	88.1	56.7	64.9	9.0	21.1	19.5	0.7	8.2	6.7	2.2	13.8	8.9
Andhra Pradesh	93.2	59.0	70.6	3.1	18.8	14.0	0.6	7.4	5.4	2.7	14.5	9.7
2009												
Coastal Andhra	62.1	66.1	89.5	0.5	2.1	1.4	0.0	0.5	0.0	0.2	3.3	2.3
Rayalaseema	29.1	55.1	61.5	1.7	17.5	17.9	0.4	5.6	5.1	0.0	17.9	15.4
Telangana	26.4	71.8	78.3	2.5	9.8	10.1	0.4	3.1	3.4	2.0	9.4	8.3
Andhra Pradesh	40.8	66.1	79.1	1.5	8.5	8.4	0.3	2.6	2.4	0.9	8.8	7.5

Source: Data compiled from Groundwater Resource Estimated Reports of different years, Department of Ground Water, GoAP.

1985, while all the command areas fell under the safe category, one was under semi-critical, two were under critical and one was under over-exploited category in non-command areas. However, the situation changed after 1993. As far as the overall groundwater development was concerned, some districts which were under safe category during earlier years slipped to semi-critical (West Godavari, Cuddapah, Ranga Reddy, Medak and Nizamabad) and critical stages (Chittoor) in 2002. In the same year, the command areas of 18 districts were under safe category, whereas non-command areas of 15 districts were under safe, four were under semi-critical, two were under critical and one was under over-exploited category, thus indicating the deteriorating situation in the non-command areas. In 2004, the overall groundwater development had pushed five districts into the semi-critical category (Chittoor, Cuddapah, Kurnool, Medak and Nizamabad). But the worrying feature in this year was that the non-command areas of as many as 10 districts were under semi-critical and three were under the critical category. Even the command areas in two districts came under the semi-critical category during the same year.

In 2007, two of the districts slipped into the semi-critical category (Medak and Nizamabad), while two slipped into the critical category (Anantapur and Ranga Reddy), as far as the overall groundwater development was concerned. However, the situation with respect to non-command areas was found to deteriorate as 12 districts were under the safe category, 8 districts were under semi-critical, while one each came under the critical and the over-exploited categories. On the other hand, the situation in command areas seems to have improved as 19 districts were under the safe category. The overall situation with respect to groundwater development seems to have worsened during 2009 when compared to 2007 as six of the districts fell under the semi-critical category (Chittoor, Anantapur, Ranga Reddy, Medak, Nizamabad and Warangal), whereas in 2007 two districts were under semi-critical and two districts were under critical category. Moreover, the number of districts falling under safe category had reduced from 18 in 2007 to 16 in 2009. While the situation in the command areas was observed to be deteriorating in 2009 when compared to 2007, it was improving in the non-command areas.

Stage of Groundwater Development by Assessment Units (Mandals)

As far as the proportion of assessment units (both command and non-command areas) falling under safe category was concerned, the share of safe assessment units at aggregate level (AP) increased in 1993 (91%) when compared to 1985 (81%), after which it declined in 2002 (75.9%) and 2004 (65.3%) and thereafter, increased in 2007 (70.6%) and in 2009 (79.1%). The proportion of assessment units (both command and non-command areas) falling under safe category in Coastal Andhra region remained the same (92%) during 1985, 1993 and 2002, after which it declined in 2004 (86.4%) and thereafter, increased in 2007 (90.2%) and again declined in 2009 (89.5%). While in 1985, the performance of Telangana Region was the least compared to the other two regions, its performance in 1993 was similar to that of Rayalaseema Region. However, from 2002 onwards, the situation in Rayalaseema Region became worse than the other two regions. The Telangana Region followed a similar trend as that of AP.

The proportion of assessment units under the safe category in the command areas in the Telangana Region was lesser when compared to other two regions, except in 2007 (data in this regard are available only after 2002 onwards). Moreover, the percentage of assessment units under the safe category in the command areas showed a declining trend in the Rayalaseema Region.

With regard to the percentage of safe category, the assessment units in the districts under the non-command areas, the trend observed in Coastal Andhra was the same as that of AP (percentage of safe assessment units declined in 2004 and again increased in 2007) except in 2009 (percentage of safe assessment units in AP was increasing whereas it was declining in Coastal Andhra). One important observation is that in 2009, the situation in Coastal Andhra Region worsened compared to other years (2002, 2004 and 2007), while in other two regions there was improvement.

The overall percentage of assessment units falling under the semi-critical category across regions was more in 2004 (14.7%) and less in 1993 (5%) at aggregate level. The overall percentage of assessment units falling under the semi-critical category in the command areas was found to be more in Telangana Region than in the other two regions. Moreover, an increasing trend was also observed in the command areas of this region up to 2007. No single assessment unit was under semi-critical category in the command areas of the Rayalaseema Region during the years 2002 and 2004. As far as the non-command areas at the aggregate level were concerned, the percentage of assessment units falling under the semi-critical category showed an increasing trend from 16.3 per cent in 2002 to 18.8 per cent in 2007. However, it had drastically declined to 8.5 per cent in 2009. The overall percentages of assessment units falling under the semi-critical category in the non-command areas were more in the Telangana Region than in the other two regions in 2002 and 2004. However, Rayalaseema Region overtook Telangana after 2007.

Assessment units (at the aggregate level for both command and non-command areas) falling under the critical category increased over the years except in 1993 and 2009. In Rayalaseema, the percentage of assessment units under the critical category at the aggregate level was relatively higher than the other two regions (lowest in Coastal Andhra Region), except 2004 and greater than the percentage of overall Andhra Pradesh. However, fluctuations were observed during these years (decreasing in 1993, 2004 and 2009; increasing in 2002 and 2007). The percentage of assessment units falling under the critical category in the command areas were more in Rayalaseema Region in 2002 and 2007 than in the other two regions. However, the situation was found to be similar in both

Rayalaseema and Telangana in 2009. No single assessment unit was under the critical category in the command areas of the Coastal Andhra Region during the years 2002, 2004, 2007 and 2009. In case of non-command areas of Coastal Andhra, the percentage of assessment units falling under the critical category increased marginally except in 2009.

At the aggregate level, the percentage of assessment units falling under the over-exploited category ranged from 0.3 per cent in 1993 to 14.7 per cent in 2004. Moreover, the overall percentage of assessment units falling under the over-exploited category were more in Rayalaseema Region compared to the other two regions (lowest in Coastal Andhra), except in 1993, and it was also higher than overall Andhra Pradesh. In 2002, no single assessment unit came under this category in the command areas across all regions. A high percentage of assessment units was recorded in 2004 for all the regions, compared to the other years, except, in the Rayalaseema Region, where higher percentage of assessment units were recorded in 2007. The trend in the non-command areas across different regions showed that Coastal Andhra performed better than the other two regions – the percentage of assessment units falling under the over-exploited category were lesser here compared to the other two regions. The Rayalaseema Region had a higher percentage of ploited units, followed by Telangana in all the estimated years.

DPAP and Non-DPAP Districts

With respect to the percentage of assessment units falling under the safe category across DPAP and non-DPAP Districts, at the aggregate level, while the performance of DPAP districts were better than that of the non-DPAP districts during 1985 and 1993, the non-DPAP districts performed better than DPAP districts during the other estimated years (2002, 2004, 2007 and 2009). In 2004, the percentage of assessment units falling under the safe category across DPAP and non-DPAP districts in non-command areas was found to be lesser compared to the other years (Table 2.10).

Table 2.10

DPAP and Non-DPAP Districts-wise Percentage of Assessment Units (Mandals) Falling under Different Categories in AP

Scheme/ Year	Categorisation of Assessment Units											
	Safe			Semi-Critical			Critical			Over-Exploited		
	C	NC	Total	C	NC	Total	C	NC	Total	C	NC	Total
DPAP												
1985	NA	NA	82.4	NA	NA	9.8	NA	NA	4.4	NA	NA	3.2
1993	NA	NA	92.9	NA	NA	5.2	NA	NA	1.4	NA	NA	0.0
2002	93.9	64.1	70.1	0.0	15.7	12.5	0.6	7.3	6.2	0.0	13.1	11.1
2004	94.5	50.4	56.3	1.2	21.0	19.8	0.0	7.2	6.0	4.3	21.0	17.7
2007	92.9	57.1	64.1	1.8	20.4	16.8	1.2	7.3	6.6	4.2	15.2	12.5
2009	28.0	67.1	74.7	0.8	13.0	12.2	0.2	4.0	3.6	0.2	10.9	9.3
Non-DPAP												
1985	NA	NA	79.9	NA	NA	13.0	NA	NA	3.6	NA	NA	3.4
1993	NA	NA	88.5	NA	NA	4.6	NA	NA	1.3	NA	NA	0.6

Cont'd...

Cont'd

2002	100	89.0	83.4	1.6	17.5	6.3	0.0	3.3	1.9	0.0	12.5	7.5
2004	91.4	56.4	77.1	5.1	11.3	8.0	0.6	8.0	4.0	2.9	23.7	10.7
2007	94.0	63.9	79.0	3.8	16.1	10.3	0.3	7.8	3.8	1.9	13.1	6.1
2009	57.7	64.8	84.9	2.5	2.5	3.4	0.4	0.8	0.8	1.9	6.1	5.0
AP												
1985	NA	NA	81.4	NA	NA	11.2	NA	NA	4.1	NA	NA	3.2
1993	NA	NA	91.1	NA	NA	5.0	NA	NA	1.4	NA	NA	0.3
2002	99.4	72.9	75.9	1.0	16.3	9.8	0.21	5.9	4.3	0.0	12.8	9.6
2004	92.5	52.5	65.3	3.8	17.6	14.7	0.42	7.5	5.1	3.4	22.0	14.7
2007	93.2	59.0	70.6	3.1	18.8	14.0	0.62	7.4	5.4	2.7	14.5	9.7
2009	40.8	66.1	79.1	1.5	8.5	8.4	0.3	2.6	2.4	0.9	8.8	7.5

Source: Data compiled from different Groundwater Resource Estimated years Reports, Ground Water Department, GoAP.

Note: 1. C-Command (*ayacut*); NC-Non Command (Non-*ayacut*); Total (both Command and Non-Command areas).

2. In 1985 and 1993 groundwater resource estimations calculated/worked out only taken the total overall-wise Groundwater Resource (both Command and Non-Command area together) and other Resource Estimations are taken in Command, Non-Command and overall-wise (2002, 2004, 2007 and 2009).

The percentage of assessment units falling under the semi-critical category under DPAP was higher than the non-DPAP districts at aggregate level in all the years except in 1985. While, in the command areas, the percentage of assessment units falling under the semi-critical category under non-DPAP was higher than in the DPAP districts in all the assessment years, in case of non-command areas, the reverse trend (the percentage of assessment units falling under the semi-critical category are more in DPAP than in the non-DPAP districts) was observed except in 2002.

Furthermore, the percentage of assessment units falling under the critical category under DPAP was higher than the non-DPAP districts at the aggregate level. While, the percentage of assessment units falling under the critical category in the non-command areas of non-DPAP districts was higher than the DPAP districts in 2004 and 2007, the reverse (the percentage of assessment units falling under the critical category are more in DPAP than in the non-DPAP districts) was found in 2002 and 2009.

Similarly, the percentage of assessment units falling under the over-exploited category in DPAP districts was higher than the non-DPAP districts, at the aggregate level, except in 1985 and 1993. And the percentage of assessment units falling under the over-exploited category in the command areas of DPAP districts was higher than the non-DPAP districts in 2004 and 2007, whereas the same was found to be more in non-DPAP districts in 2009. In case of the non-command areas, the percentage of assessment units were found to be more in DPAP than non-DPAP districts, except in 2004.

When the performances of the command and non-command areas are compared within their respective districts (DPAP and Non-DPAP districts), the command areas are found to be performing better than their non-command counterparts in all the estimation years (2002, 2004, 2007 and 2009).

The Micro Picture: Over-Exploited Villages

In AP, the number of over-exploited villages had gone up from 1,481 in 2002 to 2123 in 2009. However, fluctuations were observed during different years. This number increased to 4,190 during 2004 due to the severe drought conditions (Table 2.11). Across the regions the number of over-exploited villages went up by 86 per cent in Telangana followed by 25 per cent in Rayalaseema, whereas in Coastal Andhra the number has declined by 19 per cent over these seven years. However, these numbers are fluctuating year after year – when number of

Photo 2.1

Over-Exploitation of Groundwater: Failure of Dugwells

Table 2.11

Scheme/Region-wise Distribution of Over-Exploited (OE) Villages in OE Basins and Number of Failed Wells

Category		*No of OE Villages*				*No of Failed Wells*		*% of Assessment Units Reporting Repeated Depletion up to 2008*	*% of Assessment Units Reporting Repeated Depletion up to 2009*
		2002	*2005*	*2008*	*2009*	*2002*	*2008*		
Scheme	DPAP	1,013	2,987	2,572	1,601	2,04,323 (13,604)	1,90,638 (30,487)	27	16.93
	Non-DPAP	468	1,203	877	522	55,740 (7,633)	43,349 (9,110)	12	7.56
Region	Coastal Andhra	183	528	406	148	39,282 (6,379)	42,866 (9,401)	9.6	2.6
	Rayalaseema	726	1,405	1,446	911	72,790 (3,801)	1,00,320 (9,034)	50	23.5
	Telangana	572	2,257	1,597	1,064	1,47,991 (11,057)	90,801 (21,162)	39.8	13.2
	AP	1,481	4,190	3,449	2,123	2,60,063 (21,237)	2,33,987 (39,597)	30.3	11.3

Note: 1. Figures in the brackets indicate the total no. of failed tubewells.
2. For District-wise information see Annexure Table 2.

Source: 1. OE Villages Data (data compiled from): GoAP, Groundwater Resource Estimated Years Reports, Ground Water Department, Hyderabad.
2. Failed Well Data (data compiled from): GoAP, 3rd and 4th Minor Irrigation Census, Directorate of Economics and Statistics, Hyderabad.

over-exploited villages up to 2008 was considered, the increase was found to be almost doubled in Coastal and Rayalaseema Regions, whereas the same was almost three times in Telangana. Though these figures are based on the sample wells, they reflect the severity of groundwater depletion at the micro-level.

Among the assessment units, about 11 per cent reported the depletion repeatedly in 2009, whereas 30 per cent of the assessment units had reported repeated depletion up to 2008. As far as the proportion of repeated units are concerned, it is observed to be the highest in Rayalaseema (23.5% in 2009 and 50% in 2008) followed by Telangana (13.2% in 2009 and 39.8% in 2008) and Coastal Andhra (2.6% in 2009 and 9.6% in 2008). The average number of times repeated in this depletion category (other than safe category) was more in non-command areas when compared to command areas in all the three regions. Similarly, the percentage of assessment units repeated and the average number of times repeated were more in DPAP districts than in the non-DPAP districts. In the DPAP districts, the extents of repeated units were 17 per cent in 2009 and 27 per cent in 2008, as against 8 per cent and 12 per cent respectively during 2009 and 2008 in the non-DPAP districts.

Photo 2.2

Over-Exploitation of Groundwater: Existence of Inwell Bores

For an in-depth understanding of the severity of groundwater depletion, the relationship between 'over-exploitation' and the 'incidence of well failures' during 2002 and 2008† were examined with the help of simple correlation (Table 2.12). However, statistically significant (at 1% level) positive relationship between these two are found only in 2008 at aggregate level. When this relationship was examined region-wise for these two years, the positive and statistically significant relationship between 'over-exploitation' and the 'incidence of well failures' was found only in case of Rayalaseema region in 2008.

Table 2.12

Correlation between Over-Exploitation and the Incidence of Well Failures

Region	*Total Wells*		*Tubewells*	
	Estimated Years		*Estimated Years*	
	2002	*2008*	*2002*	*2008*
Coastal Andhra	–0.202	0.566	–0.088	0.618***
Rayalaseema	0.564	0.981**	0.099	0.708
Telangana	–0.179	0.132	0.21	0.161
Andhra Pradesh	0.278	0.692*	0.112	0.332

Note: Figures in the brackets refer to the Coefficient of Correlation and *, **, *** indicate level of significance at 1, 5, 10 per cent respectively.

Apart from examining the relationship between 'over-exploitation' and the 'incidence of well failures' (all types of wells), an attempt has also been made to examine the relationship with respect to number of failed tubewells. While the relationship between 'over-exploitation' and the 'incidence of well failures' has not turned out to be statistically significant in any of these two years at aggregate level, the relationship was

† Information related to incidence of well failures are available corresponding to these two years only.

found to be statistically significant in case of Coastal Andhra region during 2008.

Trends in Groundwater Irrigation

The area under well irrigation reflects the changes in groundwater development. Historically, the major sources of irrigation in AP are tanks, canals and wells in the same order of importance. Till the early 1970s, tanks were the dominant sources of irrigation in the Telangana and Rayalaseema Regions, while canals were the main source in the Coastal Andhra Region. After the 1970s, well irrigation emerged as the major source in Telangana and Rayalaseema Regions. Over a period of four and half decades, the proportion of area under irrigation in the state went up – from 27 per cent in 1963 to 40 per cent in 2008 (Table 2.13). The growth in the area under irrigation was found to be more in the Telangana Region (from 21% to 38% between 1963 and 2008) when compared to the Coastal and Rayalaseema Regions, resulting in a substantial decrease in regional disparities. During this period, the intra-regional disparities also decreased in all the three regions.

Across the sources, the area under canals increased by three percentage points in Coastal Andhra and one percentage point in Telangana Region, between 1963 and 2008. The Coastal Andhra Region experienced 4 per cent decline in the area under canals between 1993 and 2008, which could be due to the severe scarcity of water in the major systems during the period ending in 2008. After taking this into account, the increase in area under canal was found to be more in the Coastal Andhra Region between 1963 and 1983, which stagnated after 1983. Similarly, the picture is of stagnation or marginal improvement in the case of Rayalaseema. Inter as well as intra-regional disparities in the area under canals decreased substantially over the period of 45 years. While tank irrigation declined in all the regions, well irrigation gained more in Telangana when compared to other regions. The increase in well irrigation was the main reason for the overall decline in disparities in the state.

Telangana and Rayalaseema have experienced drastic shifts in the composition of irrigation. By 1980s, well irrigation was the dominant source of irrigation, replacing tank irrigation in the two regions. Though canal irrigation still dominates in the Coastal Andhra Region, well irrigation has replaced tank irrigation in the second place. The reasons for this are now well established (Reddy and Behera, 2009). The relative shares of the three important sources in the net irrigated area indicate that well irrigation in Telangana has gone up from 12 per cent to as much as 72 per cent, against a marginal increase in the case of canal irrigation (Table 2.13). Even the Coastal Andhra Region recorded a five-fold increase in the area under wells, while area under canals increased from 46 per cent to 55 per cent. The proportion of area under canals remained the same at 19 per cent in Rayalaseema, though it experienced an increase in well irrigation by almost three times.

Table 2.13

Source-wise Area Irrigated (Area Irrigated by Source/Net Irrigated Area) Across Regions of AP (%)

Source of Irrigation	*Region*	*Triennium Ending*					
		1963	*1973*	*1983*	*1993*	*2003*	*2008*
	Coastal Andhra	46	62	63	60	57	55
Canal	Rayalaseema	19	29	31	25	20	19
	Telangana	14	27	27	21	17	13
	AP	**12**	**13**	**16**	**16**	**15**	**16**
	Coastal Andhra	24	26	23	19	18	15
Tank	Rayalaseema	32	29	21	15	12	8
	Telangana	49	39	37	20	15	12
Well	**AP**	**11**	**8**	**9**	**7**	**6**	**5**
	Coastal Andhra	5	9	12	16	21	25
	Rayalaseema	24	37	44	57	66	72
	Telangana	12	26	32	54	64	72
	AP	**4**	**6**	**7**	**13**	**17**	**19**
	Coastal Andhra	45	50	55	57	57	55
Total	Rayalaseema	15	17	19	21	22	21
	Telangana	21	16	23	32	35	38
	AP	**27**	**28**	**33**	**37**	**40**	**40**

Source: GoAP, Data compiled from Season and Crop Reports of various years, Directorate of Economics & Statistics, Hyderabad.

The area under groundwater irrigation is nearly equal to the area under irrigation by all surface water sources put together, especially during the years of low rainfall (Figure 2.2). The population of wells increased from 0.8 million (0.7 million dugwells and 0.1 million bore wells) in 1971 to about 2.5 million (0.9 million dugwells and 1.6 million borewells) in 2007. The area under groundwater irrigation had increased from 0.8 million hectares to about 2.8 million hectares during the same period. It can be seen that the area irrigated per well is almost constant, but water is being drawn from deeper depths. The increase in exploitation of groundwater in some places is alarming and may not be sustainable unless measures are taken to control its use by increasing its efficiency.

Figure 2.2

Gross Area Irrigated by Groundwater and Surface Water (Lakh ha)

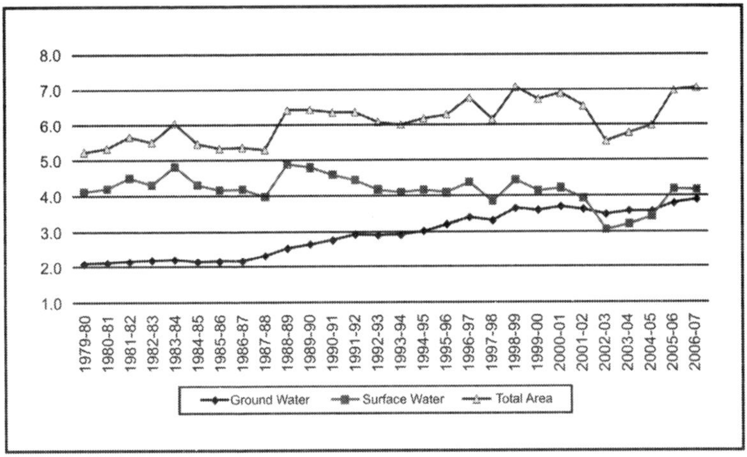

On an average, the density of wells increased from five wells to over 10 wells per sq km. However, in hard rock areas, it is over 20 wells per sq km, while in some pockets it is as high as 100 wells per sq km. Consequently, well yields decreased considerably and water levels went down alarmingly. About 48 per cent of the net groundwater availability is in command areas, which constitute about 23 per cent of the state's geographical area, and where groundwater utilisation is only 25 per cent. Here, the problem is of surplus, resulting in water logging and water quality problem.

The existing surface water bodies and canal seepages are able to contribute about 4 per cent to 5 per cent towards groundwater recharge, and about 9 per cent to 10 per cent is added by way of natural infiltration. Large scale recharge measures implemented considering riparian rights in the past could increase recharge only by 1 per cent, which also requires proper maintenance and is not very cost-effective (Kumar et al., 2008). Water levels during the pre- and post-monsoon periods (see Maps 19 & 20 in Annexure) indicate that groundwater depths are highly linked to rainfall with a little lag. Water tables had gone below 10 m range in Telangana Region during 2004–05 and 2008–09, while it increased during 2010 (Figure 2.3). Groundwater depths are the highest in the Rayalaseema Region during most of the years. Groundwater depletion adversely affects the small and marginal farmers disproportionately (Reddy, 2004). Though the small and marginal farmers are now able to invest in groundwater extraction due to availability of cheap technologies, they are often at a disadvantageous position while competing with the large farmers in well deepening. As a

Figure 2.3

Region-wise Average Depth of Water Level in Pre- and Post-Monsoon Periods

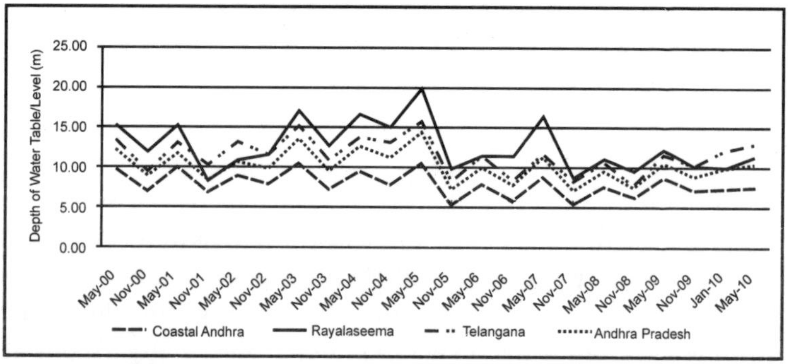

Source: GoAP, Ground Water Department, Hyderabad.

result, they become the first victims of groundwater depletion and pay huge price in terms of direct and indirect costs. Their investments become dead or waste as they are not able to compete with large farmers in investing for deeper wells.

Free Power and Groundwater Development

Groundwater development and degradation are often validly linked to the energy policies of the state. Subsidies on power (often charged at flat rate irrespective of the quantity consumed) are expected to further aggravate groundwater mining. While there are pros and cons of power subsidies on groundwater in specific and agriculture development in general, it is arguably detrimental from the resource (environment) point of view. The arguments in favour of power subsidies are that they enhance the viability of farming and increase the access to water either through water markets or otherwise, especially among the small and marginal farmers. On the other hand, it is argued that small and marginal farmers become the victims of over-exploitation (Reddy, 2004). In either case, resource degradation is imminent, resulting in 'tragedy of commons' in the long run.

The announcement of free power to farmers in 2004 by the AP Government is seen as populist act that is unmindful of economic and environmental consequences. It was argued that the policy would not only increase the power consumption adding to the burden of the exchequer, but also aggravate the problem of the already dwindling resource. Here, an attempt is made to assess the impacts of free power policy using the official data. Different indicators are examined to assess the impact of free power. These indicators include growth in agricultural service connections, energisation of wells and power consumption. None of the indicators have revealed any significant changes after the introduction of free power in 2004. The number of agricultural service connections had reached their peak during 2000–01 and declined drastically during 2004–05 (Table 2.14). A similar trend was observed in Rayalaseema and Telangana Regions. In fact, the Coastal Andhra Region recorded only a marginal increase in the number of service connections during 2004–05. This could be due to the severe drought conditions prevailing between 2001 and 2004. The impact is severe in the rain-fed regions of Telangana and Rayalaseema. Though there was substantial improvement in the number of service connections during 2006–07, the number was much below the 2000–01 peak at the aggregate level and also in Rayalaseema and Telangana Regions. Moreover, the increase was mainly due to better groundwater situation during the post-2004 period and cannot be attributed to the free power policy of the state. This is evident from the proportion of wells energised during this period and the actual power consumption.

Table 2.14

Year-wise Growth in Agricultural Service Connections Across Regions

Region	*No. of Agricultural Service Connections during the Reference Years*				
	1984–85	*1993–94*	*2000–01*	*2004–05*	*2006–07*
Coastal Andhra	8,430	20,085	22,239	22,586	25,015
Rayalaseema	4,632	23,109	26,001	15,656	17,624
Telangana	31,182	58,983	70,053	26,518	44,338
Andhra Pradesh	44,244	1,02,177	1,18,293	64,760	86,977

Source: 1. Data compiled from APTRANSCo Ltd., Power Development in Andhra Pradesh (Statistics) various years, Hyderabad.
2. GoAP Data compiled from Statistical Abstract of Andhra Pradesh various years, Directorate of Economics and Statistics, Hyderabad.

The proportion of wells energised shows a secular trend over the period of 25 years in the Coastal and Rayalaseema regions, while the Telangana Region recorded a jump in the energisation of wells after 1993–94 (Figure 2.4). By mid-1990s, Telangana overtook Rayalaseema in well energisation. In fact, the share of Telangana in well energisation was the main reason for the higher state average after mid-nineties. However, such shifts or changes were not evident after the advent of free power policy in 2004. This was also evident in the case of power consumption across regions, which show a smooth trend (Figure 2.5). The consumption of energy in the DPAP districts, which have majority of the borewells and electric pump sets,

Figure 2.4

Region-wise Proportion of Wells Energised Over the Years

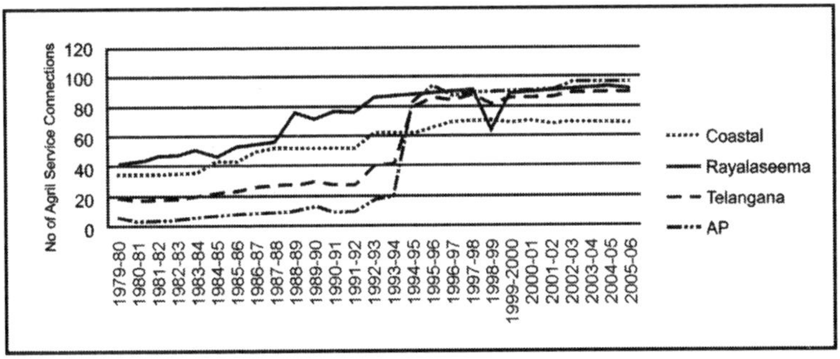

Source: 1. Data compiled from APTRANSCO Ltd. Power Development in Andhra Pradesh (Statistics) various years, Hyderabad.
2. GoAP, Data compiled from Statistical Abstract of Andhra Pradesh, various years, Directorate of Economics and Statistics, Hyderabad.

Figure 2.5

Three Year Moving Average of Electricity Consumption in Agriculture by Regions

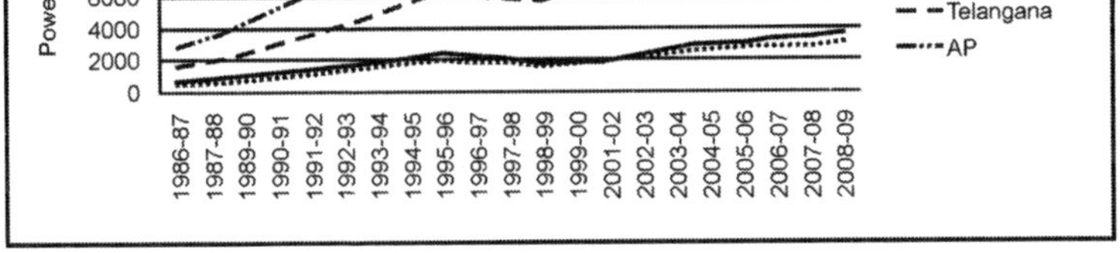

Source: 1. Data compiled from APTRANSCO Ltd. Power Development in Andhra Pradesh (Statistics) various years, Hyderabad.
2. GoAP, Data compiled from Statistical Abstract of Andhra Pradesh, various years, Directorate of Economics and Statistics, Hyderabad.

has also not shown any shift after the advent of free power policy (Figure 2.6).

Interestingly, the increasing trends observed in the case of pumps energised and the total power consumption in each year, the per pump power consumption has remained the same over

Figure 2.6

Three Year Moving Average of Energy Consumption in Agriculture by DPAP and Non-DPAP Districts

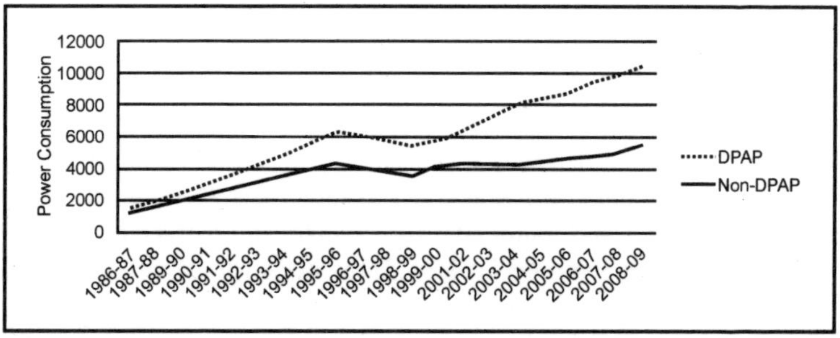

Source: 1. Data compiled from APTRANSCO Ltd. Power Development in Andhra Pradesh (Statistics), various years, Hyderabad.

2. GoAP, Data compiled from Statistical Abstract of Andhra Pradesh, various years, Directorate of Economics and Statistics, Hyderabad.

Figure 2.7

Three Year Moving Average of Energisation and Electricity Consumption in Agriculture

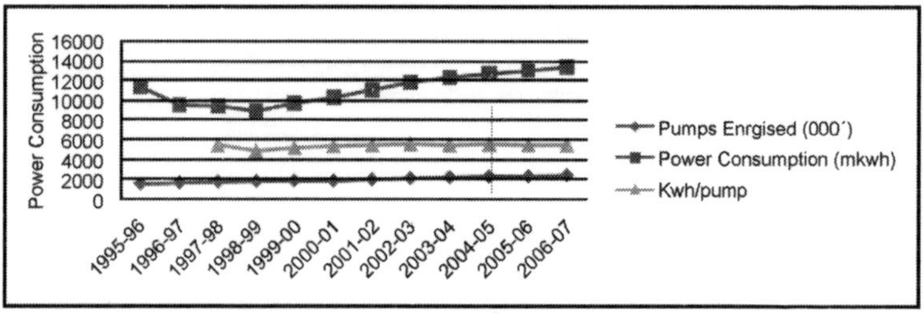

Source: 1. Data compiled from APTRANSCO Ltd. Power Development in Andhra Pradesh (Statistics), various years, Hyderabad.

2. GoAP, Data compiled from Statistical Abstract of Andhra Pradesh, various years, Directorate of Economics and Statistics, Hyderabad.

the years (Figure 2.7). This indicates that there is no change in power consumption at the individual level, before or after free power policy. This could be due to supply regulation of power at the state level. In fact, farmers complain that they get less than seven hours of supply against the promised nine hours of supply per day. As a result, the aggregate power consumption has gone up from about 1,200 million kilowatt hours in 2004–05 to about 1,350 million kilowatt hours. The financial burden of the free power policy is about ₹ 1,350 crores at the rate of ₹ 1 per kilowatt hour. However, as the data clearly indicates, this burden is not due to the free power policy. For, power was already subsidised heavily even prior to the free power policy. Under the flat rate regime, the state was collecting only about ₹ 3 crores at the rate ranging from ₹ 1,800 to ₹ 2,640 per connection respectively for 3 and 5 HP pumps.

Since there was no increase in power consumption (apart from the normal), the free power burden is mainly in terms of loss in revenue due to the flat rate collections from the existing number of energised wells. At the present level of energisation (24.48 lakh pumps in 2006–07) with an average flat rate of ₹ 2,200 per pump per year the burden on the state is ₹ 538.56 crore, which has gone up from more than ₹ 400 crore during 2004–05.

In effect, the state is losing more than ₹ 500 crore per year and this would go up over the years with increased number of energised wells. But for the supply regulation, the burden would have been higher. The state has adopted a dual policy of populism coupled with supply regulation which helped in checking the financial burden and also in maintaining the *status quo* in groundwater development. That is, the free power policy has not triggered an increased pace in the race for groundwater exploitation. The benefit to the farmers is only marginal (₹ 2,200 per pump) on an average, if not negative, considering the reduced hours of power supply. The benefits seem to be more psychological rather than real to the farming community.

Conclusions

On the whole, the spatio-temporal analysis of groundwater development has been looked at from multiple dimensions and the assessments are re-emphasised from all the angles. The analysis brings out some interesting aspects. These include:

- The methodological basis of groundwater assessment is rather weak and hence, the assessments may have limited use for the farming communities.
- There has been a secular trend in groundwater development over the years.
- This trend is only broken due to severe droughts or very good monsoons.
- Regional variations indicate that Telangana had a late entry in the case of mechanisation and energisation of groundwater exploitation, though it had overtaken the Rayalaseema Region by mid nineties.
- The level of groundwater development and its adverse impacts are more severe in Rayalaseema Region.
- There is imbalance in the development and available groundwater in the command and non-command areas. While the command areas have under-development of groundwater, the non-command areas have excess development.
- Similarly, DPAP districts face the adverse impacts of groundwater development when compared to the non-DPAP districts.
- Though the level of groundwater development at the aggregate level is not alarming, the micro-situation is a cause of concern as the number of villages included under the over-exploited category has been increasing over the years.
- The trends in groundwater development are reflected very well in the area irrigated under wells. Though the area under well irrigation is expanding, the area irrigated per well is either stagnant or declining.
- The micro-impacts are clearly seen in the case of well failures and the resulting farm distress in some regions.
- The free power policy of the state has neither helped in expanding the area under wells nor reduced the burden on the farmers substantially.
- The dual policy of free power and supply regulation does not seem to have any significant impact on agriculture.

3

Factors Influencing Groundwater Development

District Wise Analysis

At this juncture it would be pertinent to examine the factors determining the variations in groundwater development across the districts in the state. For this purpose, multiple regression analysis has been adopted using a number of indicators that influence groundwater development. The basic specification is as follows:

$$GWD_{dt} = f(NF_{dt}, EF_{dt}, SF_{dt}, DF_{dt}) + U_{dt}$$

Where,

GWD_{dt} = Groundwater Development measured in terms of extent of utilisation with reference to potential in district '*d*' at time '*t*'.

NF_{dt} = The set of Natural Factors such as rainfall, irrigation, irrigation intensity, surface water bodies, canal irrigation, cropping intensity, etc. in the district.

EF_{dt} = Economic Factors such as per capital income, extent of poverty, number of borewells, number of agricultural power connections, etc. in the district.

SF_{dt} = Social Factors such as Human Development Index, literacy level, etc. in the district.

DF_{dt} = Development Factors such as coverage under watershed development programme in the district.

U_{dt} = Error term.

The selection of independent variables is based on the theoretical considerations and the availability of data at the district level. The variables are drawn mainly from different sources such as statistical abstracts, season and crop reports, minor irrigation census, population census and departmental records including GEC Records. An exhaustive list of indicators that are likely to influence the performance was prepared. All these variables were tried in different combinations and permutations. But, some of the variables, though important, did not find place in the specifications due to various reasons including multi-collinearity, non-significance, absence of variation and also unavailability of data.* The details of variable measurement and their theoretical/expected impact on groundwater development are presented in Table 3.1.

Linear regressions applying Ordinary Least Squares (OLS) were estimated to regress the dependent variable (GWD) against the selected independent variables (SPSS package). Regressions were run on cross sectional data across the districts. Various permutations and combinations of independent variables were used to arrive at the best fits. The estimates were carried out for all the six groundwater assessment years to assess the

---

* Variables like percentage of the geographical area under cropping were tried in the regression models pertaining to all the six groundwater assessment years, but they did not turn out to be statistically significant in any of the cases. On the contrary their inclusion introduced multi-collinearity and therefore, dropped from the final specifications. Data related to evapo-transpiration was not available at the district level and hence not included in the analysis.

robustness of the estimates. Further, estimates were carried out for command/non-command and DPAP/non-DPAP areas as well, though the results are presented for non-command and DPAP areas only.

Table 3.1

Measurement and Expected Signs of the Selected Variables

Variable	*Measurement*	*Expected Impact*
Actual Annual Rainfall (AARF)	In mm per year	+ve
Irrigation Intensity (II)	Gross/net area irrigated (in %)	+ve
No. of Dugwells	Actual numbers	–/+ve
No. of Tubewells	Actual numbers	+ve
Per Capita Income (PCI)	Rupees per year	+ve
No. of Tanks	Actual numbers	–/+ve
Human Development Index (HDI)	Index of different indicators	+/–ve
Area under Canal Irrigation	In acres	–ve
Cropping Intensity (CI)	Gross/net cropped area (in %)	+ve
Area under Groundnut Crop	In acres	+ve
% of Below Poverty Line (BPL) Population	Percentage	–ve
% of Literate Population	Percentage	+ve
Number of Power Connections for Agriculture	Actual number	+ve
WSD Coverage	Area in hectares	–/+ve
Human Poverty Index (HPI)	Index	–ve
Area under Groundnut	In acres	+ve

Table 3.2

Factors Influencing Groundwater Development Over the Years in Non-Command and DPAP Areas

Variable	Year					
	1985	1993	2002	2004	2007	2009
Overall						
Irrigation Intensity (II)	NA	+	NA	+	+	NA
No. of Dugwells	NA	NS	NA	NA	NA	NA
No. of Tubewells	NA	NS	NA	+	–	NA
Per Capita Income (₹/Year)	NA	NA	NA	NA	NS	NA
No. of Tanks	NS	NA	NA	–	NA	NA
Human Development Index (HDI)	NA	+	+	NA	NA	NA
Area under Canal Irrigation (Acres)	NA	–	–	–	–	–
Cropping Intensity (CI)	NA	NA	+	NA	NA	NA
Area under Groundnut Crop (Acres)	NA	NA	+	NA	NA	NA
Actual Annual Rainfall (mm)	NA	NA	NA	+	–	NS
WSD (Area Covered in Hectares)	NA	NA	NA	NA	NA	NA
% of Below Poverty Line (BPL) Population/Human Poverty Index (HPI)*	+	NA	NA	NA	NA	–
% of Literate Population	+	NA	NA	NA	NA	NA
Number of Power Connections to Agriculture	+	NA	NA	NA	NA	+
Non-Command Areas						
Irrigation Intensity (II)	NA	+	+	NA	NA	NA
No. of Dugwells	NA	NS	NS	NA	NA	NA
No. of Tubewells	NA	NS	+	+	NA	NS
Per Capita Income (₹/Year)	NA	NA	NA	NS	NA	NA
No. of Tanks	+	NA	NA	–	NA	NA
Human Development Index	NA	+	NS	NA	NA	NA

Cont'd...

...Cont'd

Area under Canal Irrigation (Acres)	NA	–	NA	NS	–	NA
Cropping Intensity (CI)	NA	NA	NA	NA	+	NA
Actual Annual Rainfall (mm)	NA	NA	NA	–	NS	NA
WSD (Area covered in Hectares)	NA	NA	NS	–	NA	NA
% of Below Poverty Line (BPL) Population/HPI*	NA	NA	NS	NA	NA	–
% of Literate Population	+	NA	NA	NA	+	NA
Number of Power Connections to Agriculture	+	NA	NA	NA	+	+
Area under Groundnut (Acres)	–	NA	NA	NA	NA	NA
DPAP Areas						
Actual Annual Rainfall (mm)	–	–	NA	–	–	NA
Number of Power Connections to Agriculture	+	NA	NA	NA	–	NA
Area under Canal Irrigation	–	–	–	–	–	–
Cropping Intensity (CI)	NA	+	+	NA	NA	+
WSD (Area covered in Hectares)	NA	NA	+	NA	NA	+
% of Below Poverty Line (BPL) Population	NA	NA	NS	NA	NA	NA
% of Literate Population	NA	+	NA	NA	NA	NA
Irrigation Intensity (II)	NA	NA	NA	NS	NA	NA
Human Poverty Index (HPI)	NA	NA	NA	–	NA	NA

Note: + indicates positively significant; – indicates negatively significant; NS= Not Significant; NA= Not Applicable (not used in the specification). Detailed estimates are presented in the *Annexure* Table 3 to 15.

Multi-collinearity between the independent variables was checked using the Variance Inflation Factor (VIF) statistic. Multi-collinearity is not a serious problem as long as the value of VIF is below 2. The best fit specification was selected for the purpose of final analysis for each dependent variable. The indicative results are presented in Table 3.2 while the detailed results along with the descriptive statistics are presented in the *Annexure* (Table 3 to 15).

The estimates indicate that not many variables turned out to be significant across the years in both non-command as well as DPAP areas, though the selected specifications explain more than 60 per cent of the variations in the case of non-command areas except 2009 and more than 70 per cent of the variations in the case of DPAP districts except 1985 (see Annexure Table 9 to 15). Most of the indicators have shown up with expected signs. At the state level, the area under canal irrigation turned out to be significant in five of the six years, with a consistent negative sign indicating that groundwater development is limited in the canal irrigated areas (Table 3.2). On the other hand, cropping intensity has shown a positive impact on groundwater development in one year. That is, groundwater is used more intensively as the area under second and third crops increase. Agricultural service/power connection turned out to be significant in two of the years with a positive sign indicating enhancement of groundwater exploitation as a result of more number of power connections to agricultural sector. Similarly, Human Development Index (HDI) turned out to be significant in two of the years with a positive sign. This means that human development could increase groundwater development due to the overall comprehensive development reflected in the HDI. Actual rainfall was seen to have a negative impact in 2002, 2004 and 2007. But, number of tubewells also did not have a consistent sign which turned out to be significant in 2004 and 2007. It may be inferred that number of tubewells would increase exploitation of groundwater in drought conditions (2004) while in good rainfall year (2007), the demand would go down coupled with increased supply, resulting in a net negative impact on groundwater development.

In the case of non-command areas, the variables' area under canal irrigation and irrigation intensity turned out to be significant in two of the years with signs similar to that of the state level. The number of tubewells, power connections, literacy and HDI revealed a positive impact on groundwater development. In the non-command areas, all these indicators, except the area under canal irrigation, promote groundwater exploitation. The number of tanks was seen to have a positive impact in one year and a negative impact in the other year. Furthermore, it is seen that the actual rainfall and the area covered under watershed development also have a negative impact on groundwater development. That is, in the non-command areas, watershed development could lead to checking of groundwater exploitation.

However, in the DPAP districts, WSD leads to exploitation, which is also evident at the field level studies (Reddy et al., 2010). That is, groundwater use in terms of the number of wells tends to increase after the advent of watershed development. Often this results in upsetting the recharge impact of watershed development. The actual rainfall and area under canal irrigation have a more clear negative impact on groundwater development in the DPAP districts, as they turned out to be significant in most of the years. On the other hand, cropping intensity has shown a consistent positive impact in three of the years. And literacy, also in line with non-command areas, and state-level impacts, has a positive impact on groundwater development.

Overall, the regression analysis of the factors influencing groundwater development does not prove to be of much help in a better understanding of groundwater management. This is because the policy variables such as WSD, literacy and HDI did not reveal any clear impact towards checking groundwater development. This could be due to the reason that in the absence of groundwater institutions these factors may have limited influence. Besides, our analysis also does not include any institutional variables due to the non-existence of any formal groundwater institutions at the district level. Therefore, it would be pertinent to examine the impact of institutions on groundwater management that are prevalent in some of the districts. The following chapters examine these aspects in detail, on the backdrop of Andhra Pradesh.

PART II

Institutionalising Groundwater Management: A Tale of Three Participatory Models in Andhra Pradesh

4

Managing Groundwater

Review of Approaches

Introduction

While groundwater is studied extensively in terms of its hydro-geology and socio-economic aspects, sustainable management of groundwater has not been dealt with comprehensively either by researchers or policy makers. The increasing groundwater crisis consequent to its over-exploitation and degradation makes groundwater management imperative from the ecological as well as socio-economic point of view. Though the Approach Paper to the 12th Plan recognises this importance, it fails to provide any plan of action due to the absence of any clear understanding of groundwater management. The main bottleneck for bringing groundwater under a management regime is that groundwater is treated as private property by individuals, as a right attached to land ownership. Attempts towards changing this practice are not only perceived to be associated with huge transaction costs, but also resulted in socio-economic conflicts due to the existing inequity in groundwater distribution as well as its economic value.

From the economic point of view, groundwater irrigation is observed to be twice as efficient as surface water irrigation in hydrological terms (m^3/ha), and ten times preferable (Llamas and Martínez-Santos, 2005). Besides, it has a large number of *in situ* services including environmental, and is promoted as a plausible option for poverty reduction (Burke et al., 1999; Polak, 2004). Kumar (2007), for instance, estimated that the surplus value product generated from the groundwater in India's irrigated lands (15 major states) contributes nearly 5 per cent of its gross domestic product. A large fraction of the population directly or indirectly relies on groundwater resources for livelihood, as more than 60 per cent of irrigated agriculture is dependent on it (The World Bank, 2010). Groundwater plays a major role in achieving India's food security, besides turning into a net exporter of food, despite a twofold increase in population during the last 50 years (Shah, 2004). Groundwater development requires relatively smaller investment and shorter implementation periods when compared to the traditional surface irrigation system (Valencia Statement, 2004).

These virtues of groundwater in the absence of clearly defined property rights have resulted in the sharp increase in groundwater use, and over-exploitation as well as degradation of the resource (Dhawan, 1995; Moench, 1992; Bhatia, 1992). In India, the declining groundwater table has resulted in increasing the cost of pumping with declining yield. Failure of wells has become a common phenomenon in recent years, and has been causing widespread farmer distress (Reddy and Galab, 2006). Overdraft is generally a by-product of population growth, economic expansion, distorting impacts of subsidies, and financial incentives, in addition to the spread of energised pumping technologies (Burke et al., 1999). According to Shah et al. (2000), groundwater development faces challenges due to three major problems: depletion due to overdraft, insufficient conjunctive use, and pollution due to growing agricultural activities.

About a quarter of India's agricultural production has been at risk due to growing depletion (Shah et al., 2000), which results in the persistence of poverty and low growth – a situation that has further aggravated during recent years. Intensive use of fertilisers and pesticides, leaching from compost pits, animal refuse, dumping grounds for garbage, seepage from septic tanks and sewage, etc. affect groundwater quality (Burke et al., 1999; Sharma, 2009). Another serious issue of groundwater quality is arsenic – approximately 50 million people worldwide are affected by arsenic (Alaerts and Khoury, 2004).

Groundwater over-exploitation thus has serious implications for achieving the Millennium Development Goals (MDGs) (The World Bank, 2010). This is because declining access to groundwater not only affects agricultural production, but also education, health, gender, child mortality, poverty and hunger (Sharma, 2009). Although groundwater is not a scarce resource in most regions, sustainable management of the resource is the crux of the problem (Burke et al., 1999). This part (II) attempts to explore the possible options for groundwater management in the Indian context. The main focus of the study is to understand the functioning and efficacy of groundwater management institutions by comparing and contrasting three participatory groundwater models in AP.

This chapter reviews the existing groundwater management practices at the policy level across the countries. Groundwater is a typical resource, as it has the attributes of common pool resource with greater feasibility for private access and management. In most situations, it is considered as a Common Property Resource (CPR) with extremely high use value (Burke, 1999). At the same time, the linkages between groundwater and land ownership facilitates private access and management. This dichotomy of common as well as private good qualities makes sustainable management of the resource extremely difficult. In some countries like Indonesia, Australia, USA and Peru, it is considered as a public good either through legal tradition or through the suppression of private ownership rights (IRM and ED, 2008). However, in countries like India, groundwater is treated as a *de facto* private property, though other precious resources, such as minerals, lying beneath private lands are treated as state property (Singh, 1995). This often results in over-exploitation of groundwater and inequity in access. In order to ensure equity and sustainable use, countries like South Africa have abolished riparian laws through delinking land and water rights (Reddy, 2007).

Lack of clarity regarding property rights on groundwater also results in the poor implementation of sanctioning and enforcing water allocation mechanisms at the policy level. Due to rigid and static governance structures, the policy makers fail to understand the changing groundwater scenario. Lack of information at appropriate scale is a bottleneck at the community level for adopting informed groundwater management practices (Reddy et al., 2011). In the absence of appropriate information coupled with high economic value, the highly heterogeneous nature of groundwater availability in space and time is turning groundwater extraction into a high-risk venture. Therefore, it is necessary to understand the existing groundwater management systems at different levels (national, state and community). Based on a review of existing literature the basic management principles being adopted for groundwater management can be broadly grouped under three approaches, viz. regulatory, economic, and community-based. In what follows, we briefly discuss these approaches.

Regulatory Approaches

Regulation is the most commonly used instrument for managing groundwater use. Regulation mechanisms include restrictions on digging new wells, well depths and the volume pumped, demarcating groundwater protection zones, etc. which are generally enforced by the state administrative process (Shah, 2009). Apart from direct regulation, indirect regulation through restricted supply of electricity for pumping, restrictions on financing, etc. are also used to manage groundwater. These regulations consist of a complex and multilayer framework of a range of constitutional and statutory provisions at the central and state levels. Groundwater management in India falls within the jurisdiction of the State Government that is responsible for the financing, cost recovery and management of all water resources (Saleth, 2005). However, the Central Government has the concurrent power to make laws with respect to any matter for any part of the territory of India.

The Indian Easements Act of 1882, which mentions the private property rights over groundwater use, forms the basis for groundwater regulation in India (Saleth, 2005). It is adopted from the English Common Law, which gives every owner of land 'the right to collect and dispose within his own limits of all water under the land which does not pass in a defined channel' (The World Bank, 2010). Thus, groundwater is treated as an appendage to land because it is an easement connected to land, and persons who own the land also own the groundwater beneath it. They have also the right to transfer rights over groundwater along with the ownership of land.

The GoI introduced a Model Groundwater Bill during 1972 constituting a groundwater management agency at the state level, which is responsible for registrations and control of larger groundwater users. Some of the major elements of this bill include power to notify areas for control and regulation of groundwater development, grant of permission to extract and use water in the notified areas, registration of existing users in the notified areas, prohibition of carrying on sinking wells, etc. The Model Groundwater (Control and Regulation) Bill of 1992 proposes a kind of groundwater permits system. However, it did not set any withdrawal limits (GoI, 1992) and is confined only to the states of Tamil Nadu, Maharashtra, and Karnataka. The National Water Policy of 2002 also makes certain provisions on the control of groundwater extraction.

During the late 1990s, AP, Tamil Nadu, and Maharashtra enacted groundwater legislations. These legislations imposed restrictions on groundwater exploitation by making registration of wells as well as rigging technologies mandatory. The implementation and enforcement of these legislations are yet to bear fruit due to various reasons. For, these legislations have failed to take spatial distribution of the resource into account by putting all the regions together, irrespective of their level of groundwater development. That is, top-down regulations take an aggregate view of the situation, and often fail to capture the local-specific conditions such as geo-hydrology and socio-economic aspects of groundwater use. Hence, they are the least likely to be socially and politically viable. Similarly, socio-economic equity is not taken into account while enforcing the regulations; i.e. treating those having and those not having wells equally. The doctrine of prior appropriation reinforced the access rights of the existing well owners while curtailing new wells in over-exploited areas. Due to the negligence and conflict of interests of all sections of the society, enforcement has received scant attention (Sharma, 1995). Similarly, the monitoring mechanism to ensure that a particular regulation is enforced is a costly and difficult task in vast and remote regions (Kumar, 2007).

Limiting the power supply and formal credit are the indirect ways of regulating groundwater use. A number of states in India follow power supply regulation for one reason or the other. The main reason, often made explicit for restricted power supply, is supply constraint as well as reducing the burden on the exchequer due to subsidised or free power supplied to the farm sector. The externality of restricted power supply is the regulation of groundwater use. In fact, farmers express, 'but for the limited power supply, their borewells would have gone dry', especially during drought years. The power supply restrictions are usually associated with subsidies or free power. The Gujarat Electricity Board (GEB) does not provide new electricity connection for extraction of groundwater in over-exploited, critical and saline areas without the consent of the Central Ground Water Authority (CGWA). It has also launched the *Jyoti Gram Scheme* (JGS), which puts separate feeders for agriculture and domestic services (Lakhina, 2007). Restricted power supply is being followed in a number of states including AP, Gujarat, etc. The power is supplied for only eight hours per day for agricultural purposes; AP has been supplying power supply for 7–9 hours a day along with the free power policy for the last 7 years. Restricted power supply policy was observed to have little consequence in the case of large pumps and multiple wells, as the effectiveness of regulations undermines not only the availability of the diesel pump-set option but also by the presence of a 'kink' in the farmers' power demands (Saleth, 2005). As a result, misuse of power as well as groundwater is widespread, as farmers leave their pumps on round the clock. Hence, the combined impact of free but limited power supply for groundwater use needs to be assessed critically.

The National Bank for Agriculture Rural Development (NABARD) has adopted a policy not to provide refinance in critical and over-exploited areas. NABARD has prescribed spacing norms for different types of areas whereby the minimum distance between two groundwater abstraction structures can be indicated (IRM and ED, 2008). According to the NABARD regulation, the farmers do not get credit for a new borewell if it is located within 200m radius of an existing borewell. Such restrictions are also imposed by other nationalised banks. Field research has shown that credit regulation was not very effective due to the availability of other credit avenues (mainly informal sources) at the village level (Kumar, 2007). This is despite the fact that the cost of credit from informal sources is high. The credit rationing policy of the banks is also trying to curb new power connections to borewells and place restrictions on electric power supply. Besides, enforcement is also lax due to the pressure on banks to achieve targets.

The Punjab Government has recently introduced the Punjab Preservation of Sub-Soil Water Ordinance 2008, which prohibits the planting of paddy by the farmers in the state before June 10, in order to conserve groundwater. The ordinance provides for the government agencies to plough the area with the standing crop of such farmers who transplant paddy before the notified date. The effectiveness of this order dissuading farmers from sowing early paddy, thereby conserving groundwater is, however, yet to be seen.

A model bill to regulate and control development of groundwater has been circulated by the Ministry of Water Resources (MoWR) to all the States/Union Territories (UTs). So far, 11 States/UTs including AP, Goa, Tamil Nadu, Kerala, West Bengal, Bihar, Himachal Pradesh, Chandigarh, Lakshadweep,

Puducherry, and Dadra and Nagar Haveli have enacted and implemented groundwater legislation. However, the effectiveness of their implementation and enforcement is not known.

Some success in reducing groundwater draft through regulatory measures have reportedly been made in a few water-scarce countries such as Jordan, where a quasi-water policy requires measuring withdrawals from the irrigation wells, enforcement of pumping quotas and levy of volumetric groundwater fee (The World Bank, 2000). However, the situation is more complex in countries such as India where millions of individual private tubewell owners, dispersed through the length and breadth of the country with varying groundwater availability and demand conditions, are engaged in groundwater extraction. Putting into effect such an approach and overseeing its implementation in a country of the size of India is nearly impossible. For, the number of groundwater structures in India is estimated at about 23–25 million (The World Bank, 2010). Maharashtra has recently developed a groundwater management model, which involves regulation of more than 1.5 million irrigation wells. It includes a levy on groundwater use and a ban on deep tubewells. The Chinese, with stronger state commitment to groundwater regulation, with a more elaborate reach and local authority structures still find it impossible to regulate groundwater overdraft in North China plains (Shah, Giordano and Wang, 2004a). Neither have the Americans been able to implement real groundwater demand management, with their elaborate structure or water rights and groundwater districts, nor the Spaniards and Mexicans, with their efforts to promote groundwater user associations.

Economic

Pricing of water or a complementary input such as electricity or diesel, water markets, and tradable water rights are some of the important economic instruments that are used in the case of groundwater management. Economic instruments include charges and taxes levied on irrigation wells or volume of water withdrawn such as the 1994 Water Law in China (Wang et al., 2007), Law of the Nation's Water in Mexico (Shah et al., 2004a; Scott and Shah. 2004; Sandoval, 2004), and Israel (Feitelson, 2006). An example of taxes as an economic instrument is found in Chennai (Briscoe, 1999). Municipal water utility is paying the farmers to sell borewell supplies in order to meet the drinking water demand in the urban areas, which created an incentive for the farmers to put water to a higher-value use and reducing mismanagement in groundwater allocation. However, it is very difficult to collect and enforce such a fee in case of large resource users or poor governance environment (Shah, 2009).

Electricity pricing is a more commonly followed instrument in India. Electricity has the potential to regulate the use of groundwater. For instance, it is argued in the context of different regions of India that *pro rata* electricity pricing enhances groundwater use efficiency and sustainability without affecting net returns from farming (Kumar, 2005; Kumar et al., 2011). The study estimates the levels of pricing at which demand for electricity and groundwater becomes elastic and shows that pricing is socio-economically viable. Further, water productivity impacts of pricing would be the highest when water is volumetrically allocated with rationing. Therefore, an effective power tariff policy, followed by the enforcement of volumetric water allocation could address the issue of efficiency, sustainability and equity in groundwater use in India (Kumar, 2005). Similarly, in the context of AP, which is the front-runner in the provision of free power along with supply restrictions, it is argued that pricing of electricity for irrigation is the only option for addressing agrarian distress (Kumar et al., 2011). However, the impact of pricing on groundwater management could also vary, depending on the water productivity (Malik, nd).

In case of diesel pricing, it was found that price rise may not necessarily result in the reduction in groundwater use (Shah, 2007). On the contrary, farmers may opt for highly water intensive and remunerative crops. However, the main difficulty with the price mechanism is that of implementation. There is lack of required administrative resources for metering and monitoring groundwater use and collecting user fees. During the 1970s, the GoI had faced difficulty in metering about 2 million wells and thus implemented a flat tariff on electricity used by agriculture. At present, the number of wells is over 20 million, aggravating administrative difficulties and transaction costs. Besides, pricing is a politically sensitive issue, especially when populism has become the norm (Kemper, 2007).

The development of private groundwater market has a long history in rural India (Pant, 2005; Saleth, 1994). Even though selling of water was traced out during the 1920s, it was only in the 1960s that systematic information started flowing (Saleth, 2005). Groundwater markets are widespread in Gujarat, Tamil Nadu, AP, Uttar Pradesh (UP), and West Bengal (IRM and ED, 2008). However, there are no clear-cut statistics about the total area under private groundwater market. Based on his studies from Gujarat and UP, Shah (1993) projected that the area irrigated under groundwater markets was about 50 per cent of the total Gross Irrigated Area (GIA) under private lift irrigation. Whereas Shankar (1992) mentions that the actual GIA ranges from 80 per cent in Gujarat to 60 per cent in UP. A Tamil Nadu study shows that it is not more than 30 per cent (Janakarajan, 1993).

A market is basically formed through a mutual understanding between two adjacent farmers to share water (Mukherji, 2007). It serves two purposes: promoting efficient use, and providing water to poor farmers who are either unable to afford wells or find it uneconomical to do so (Shah, 1989, 1993; IRM and ED, 2008). The markets also increase cropping intensity and demand for agricultural labour, which ultimately benefits the landless and wage labour (Fujita and Hussain, 1995).

The impact of water markets on groundwater demand is not necessarily negative. Though markets encourage groundwater use efficiency, they often expand the area per well due to the incentive to sell water. However, the extent of the impact again may depend on water productivity. Groundwater market in Gujarat, for instance, does not consider the limit of the resources and is thus not sustainable in the long run (Kemper, 2007). Topography and distance between the source and the field also influences sustainability. In hard rock, deep alluvial or scanty rainfall areas, development of market sharing results in over-pumping and over-exploitation (Roy, 1989). On the other hand, Shah (2009) mentions that tradable property ownership creates incentives for improving productivity and conservation.

Community Management

Community management of groundwater is very limited in its spread despite the fact that the community management of irrigation (through tanks or canals) is very old. This is mainly because groundwater resource is considered to be private property. Participatory approach to groundwater management in India is based on the Western United States' experience of the communities in aquifer management. This model was also tried in Spain and Mexico where users are registered and organised into associations with a mandate to manage sustainably (Villarroya and Aldwell, 1998; Sandoval, 2004). Thus 'community management' implies creation of self-governing water user organisations who take the responsibility of sustainable management of aquifers through collective action (Shah, 2009). The main objectives of the management process are to focus on the demand side through participatory data collection, analysis and dissemination (GoAP, 2007). It can also involve any mix of instruments including regulation, property rights, and pricing (The World Bank, 2010).

The Government of Andhra Pradesh (GoAP) and Punjab for instance started Groundwater Management Projects where farmers are equipped with the necessary data, skills and knowledge, for managing groundwater in a sustainable manner through managing and monitoring their own demand. They measure, and keep a daily track of rainfall, water levels, and well yields, calculating groundwater recharge from monsoonal rainfall, and estimating their annual water use based on the planned cropping pattern. Empirical studies show that in the years when water availability was low either due to low rainfall or high groundwater abstraction during the preceding crop season, the farmers are now able to achieve a combination of crop diversification and water-saving irrigation methods (The World Bank, 2010).

A different type of peoples' participation was observed in Rajasthan. The villagers decided to stop sinking of borewells in order to preserve and judiciously use the water resources at their disposal. As a result, no borewell is found within the 4 km radius of the village (IRM and ED, 2008). In Kerala, two community managed groundwater projects were implemented for proper utilisation of water for irrigation. As per the instructions, two persons can irrigate their land at a time. The farmers bear the electrical and Operation and Maintenance (O and M) charges and succeeded to achieve financial and source sustainability. Check Dam Movement was started in Gujarat, where farmers formed village-level local institutions (Gandhi and Sharma, 2009). Under this system, the villagers undertake planning, finance and construction of a system to check dams in and around the village in order to collect and store rainwater, recharge the groundwater aquifers, and thereby recharge the dugwells. As a result, the water table has increased, improving the agricultural income. However, there was no collective action on reducing over-extraction. The communities were self-interested and every farmer in the community was free to extract whatever they wanted, rather than focusing on collective targets for crop diversification or water use reduction.

Community-based management programmes should be designed with a shared focus on improving agricultural productivity, income and water conservation. Water use reductions should not be explicitly sought, but realised by aligning efficient irrigation interventions with farmer incentives for higher profits. The Planning Commission (2007) also agrees with the fact that community management or control would not work well unless it serves some basic needs of the farmers.

According to The World Bank (2010), stakeholders' participation in the management process is necessary because it disseminates understanding of issues that can be the impetus for up-scaling good practices for the sustainable use of groundwater. It also improves the self-regulatory capacity, counteracts corruption, and facilitates the coordination of decisions relating to groundwater, land use, and waste management. According to Burke et al. (1999), socio-economic, political and institutional factors are the main determinants, which incentivise these stakeholders for sustainable groundwater management. As reliability of water supply declines, it poses tremendous risk to the people depending on it. It also influences farmers' decisions about investment in fertiliser, seed, and other inputs; the Government and other institutional investments; and economic returns. Thus, a detailed account about how people are using groundwater, why extraction rate is tremendously increasing, the pricing mechanism and sharing structure, etc. need to be analysed for a better policy framework that bridges the gap between physical availability to administrative and institutional responses towards a sustainable management process. This calls for proper understanding of the property rights regime under which groundwater development and management falls.

There are a few participatory groundwater management initiatives implemented by different NGOs in various states (GoI, 2011). These include:

1. The APFAMGS programme in AP aimed at involving farmers in hydrologic data generation, analysis and decision-making, particularly around crop-water budgeting;
2. groundwater sharing under the Andhra Pradesh Drought Adaptation Initiative (APDAI) involving Watershed Support Services and Activities Network (WASSAN), in parts of AP;
3. experiences from Barefoot College, Tilonia, with a water budgeting tool known as *Jal Chitra*;
4. efforts by the Foundation for Ecological Security (FES) at taking a micro-watershed unit for water balance and

planning groundwater use along with communities at their sites in Rajasthan, Madhya Pradesh (MP) and AP;

5. experiences of the Advanced Centre for Water Resources Development and Management (ACWADAM) with *Samaj Pragati Sahyog* (SPS) in Bagli, MP, and with the *Pani Panchayats* in Maharashtra on knowledge-based, typology-driven, aquifer-management strategies;
6. the *Hivre Bazar* model of watershed development (WSD) and social regulation to manage water resources; and
7. social regulation of groundwater use initiated by the Centre for World Solidarity (CWS).

The operational modalities and their functioning need to be assessed critically in order to draw lessons for broader policy formulations.

Amidst this backdrop, the following chapter describes three institutional models that are addressing groundwater management in AP following different approaches. These institutions are assessed in terms of their structure, operational principles, functioning, and effectiveness in managing groundwater at the community level. A comparative assessment of the strengths and weaknesses of these approaches will be taken up in order to arrive at a feasible or acceptable institutional model for scaling up.

5

Participatory Groundwater Management

Three Approaches

Background

Water is a State Subject and so is its development, utilisation and monitoring. The Government of AP is responsible for water resource planning, storage as well as use. Several Government Departments/Agencies, NGOs and people's institutions are involved in water development, use, monitoring and regulation. Water management is encouraged through institutional arrangements such as Water User Associations (WUAs) and Tank Management Committees (TMCs). These state promoted institutional arrangements are limited to surface water resources such as canals and tanks leaving groundwater development and management to private individuals. Though effectiveness and sustainability of canal and tank management institutions are being debated (Reddy and Reddy, 2005), the need for bringing groundwater under common resource management cannot be undermined. Hitherto groundwater management is left to private individuals, as it is perceived to have high transaction cost of organising individual farmers at a scale to attain the benefits of community management.

On the other hand, as observed in the earlier section on review, there appears to be some small-scale institutional innovations that are working towards sustainable management of groundwater in different corners of the country. However, these innovations are confined to small areas in the absence of policy support in bringing groundwater under the management regime, and the possibilities for scaling up these models have not been explored. Here we make an attempt to explore the possibilities for scaling up and drawing lessons from PGM by comparing three such models that are in operation in AP.

The State of AP has a long history of community groundwater management, and was one of the first states to initiate a joint well programme way back in 1987. The three models selected are:

1. The Andhra Pradesh Farmer Managed Groundwater Systems (APFAMGS) Project , which has its origins in Andhra Pradesh Groundwater Borewell Irrigation Scheme (APWELL) programme;
2. Social regulations in Water Management (SRWM) by the CWS (NGO) and its partners; and
3. Collectivisation of borewells under the Andhra Pradesh Drought Adaptation Initiatives (APDAI) programme being implemented by WASSAN with its partner NGOs. These initiatives have different origins and approaches to PGM (Table 5.1).

All the three models have been initiated in the arid and semi-arid districts of AP, where the extent of groundwater development is quite high. In what follows we discuss these three models in detail.

Table 5.1

Groundwater Management Programmes/Project Models in Andhra Pradesh

CBGWM Model	*Description*
APFAMGS (APWELL)	(a) Dug new borewells for a group of HHs not having access to water, with clear sharing, groundwater monitoring, and water use efficiency measures. (b) Limited to 'new unexploited' areas. APWELL has been transformed into the largest groundwater awareness programme in the state premised on: (i) communities monitoring the groundwater status regularly with knowledge and scientific principles;

Cont'd...

...Cont'd

	(ii) sharing the knowledge of various alternate crop systems and evolving norms for groundwater management (with facilitation); this process will lead to lesser groundwater depletion and better management.
Social Regulations in Water Management (CWS & Partners Programme)	This programme was initiated on a limited scale and based on regulations: (i) the community adopts a norm of 'no new borewells'; (ii) increasing system efficiency through the provision of collective sprinkler irrigation sets; and (iii) borewell owners share their water with neighbouring farmers leading to substantial reduction of the number of water-less families in the village.
Collectivisation of borewells: APDAI (of DoRD, facilitated by WASSAN)	This initiative followed an 'area approach' for groundwater management where the borewell owners pool their individual borewells to provide supplemental/critical irrigation to a larger rain-fed area (entire block) for survival of rain-fed crops. The community has to abide by the following rules: (i) no new borewells for at least 10 years; (ii) all the land within the specified area (including water-less) will have a right for supplemental irrigation for Kharif rain-fed crops; and (iii) pipeline network is provided by the project so that water can be taken to any part in the block/area.

Source: Field Observation (PRA/FGD methods).

APFAMGS

The Andhra Pradesh Farmer Managed Groundwater Systems (APFAMGS) has its origin in the APWELL Project initiated by the GoI in 1987. The APWELL Project was conceived in collaboration with the Netherlands Government, which funded a number of minor irrigation schemes in AP. The APWELL Project was approved for financing by the Netherlands Government in June 1994. From April 1995 to March 2003, the

APWELL Project was implemented in seven districts of AP, viz. Prakasam, Mahbubnagar, Nalgonda, Anantapur, Kurnool, Chittoor and Cuddapah. The project was co-financed with 15 per cent (of total cost excluding establishment costs) contribution from the farmers, and the rest as a grant from the Royal Netherlands Government. The establishment costs and part of the cost of electricity infrastructure were borne by the GoI/GoAP.* Physical activities such as groundwater prospecting, drilling, yield testing, and construction of the distribution systems, were done through the Andhra Pradesh State Irrigation Development Corporation (APSIDC), with its technical staff under the Executive Engineer in each district.**

Under this programme, a total of 4,480 borewells were drilled. Of these, 3,462 were successful with yield above 1,500 gph, at 77 per cent success rate† in 470 villages, covering about 14,000 households across seven districts (Table 5.2). In the APWELL Project, farmers own and maintain the borewell irrigation systems constructed as part of the project. Under each well, the farmers formed Water User Groups (WUGs) for construction, operation, and maintenance of the borewell systems. Women WUG members formed Self-Help Groups (SHGs) for thrift and credit activities, and gradually initiated land and water-based agriculture and other supplementary income generating activities. On an average each well/WUG has four households covering 10 acres of land, i.e. an average of 2.5 acres. The average cost of a well

---

* For details on funding pattern, see APWELL Final Report. ARCADIS Euroconsult, 2003, pp. 45–50.

** A technical assistance team, consisting of national and international experts on various disciplines, based in Hyderabad, advised and coordinated project activities in the field. In each district, local NGOs were contracted to implement the social, institutional, gender, agricultural, and watershed aspects of the project. For this the NGOs appointed a dedicated team consisting of Agricultural Production Trainers (APTs), Gender Development Organisers (GDOs), Watershed Development Facilitators (WDFs), and Community Organisers (COs). A District Field Coordinator (DFC), who was part of the consultant's team, supervised the work in each district.

† About 75 per cent is the acceptable success rate.

Table 5.2

APWELL Project Coverage on Completion (up to March 2003)

District	*Villages*	*No. of Wells/ WUGs*	*Total No. of HHs*	*HHs/ WUGs*	*Total Ayacut (acres)*	*Ayacut per WUG (acres)*	*Ayacut per HHs (acres)*	*Avg. Yield (kg/ha)*	*Avg. Cost per Borewell (₹)*	*WUG Contribution/ Well (₹)*
Anantapur	39	415	1,396	3.4	4,410	10.6	3.2	4,009	1,31,724	16,159
Chittoor	110	419	2,076	5	3,481	8.3	1.7	3,109	1,41,242	17,171
Cuddapah	59	415	2,160	5.2	3,978	9.6	1.8	2,995	1,50,625	18,167
Kurnool	78	518	2,013	3.9	5,299	10.2	2.6	4,557	1,43,036	16,765
Mahbubnagar	55	821	2,741	3.3	8,605	10.5	3.1	2,604	1,29,987	15,610
Nalgonda	42	299	1,439	4.8	3,018	10.1	2.1	3,569	1,53,300	18,796
Prakasam	87	575	2,053	3.6	5,698	9.9	2.8	3,523	1,42,660	16,635
Total	470	3,462	13,878	4	34,489	10	2.5	3,523	1,40,102	19,790

Source: APWELL Project: Final Report, ARCADIS Euroconsult, 2003, pp. 76–77.

was about ₹ 1.4 lakhs of which about 14 per cent was contributed by the farmers (Table 5.2). Clusters of WUGs were formed into Borewell User Associations (BUAs), which in due course were legally registered, for training, conflict resolution, procuring agricultural inputs, marketing, agro-processing, and groundwater management. Important components of the project included: groundwater resources development where feasible, land and water management by the users, extension and training, activities for gender integration, environment management, and monitoring and evaluation.

The project clearly demonstrated that PGM is a viable concept if introduced in conjunction with groundwater development, agricultural production, institutional development, and capacity building of farming communities. The implementation process followed by the APWELL Project achieved certain important results, which are good lessons for future projects:

- The intense community organisation efforts to form and nurture WUGs assured the involvement of the farmers from the very inception of the project activities within the village.
- Compulsory inclusion of women as members of WUGs and forming SHGs helped to mainstream women farmers into the management of groundwater systems.
- At the end of project implementation, the assets created were handed over to the WUGs. Thus, the project had a distinct exit policy woven into its concept.
- The WUGs contributed 14 per cent of the cost (excluding administrative charges). This ensured greater sense of ownership among them.
- Every member of the WUG was given a pipe outlet on his/her land, assuring equity in water distribution and reducing water conveyance loss.
- Intensive capacity building through training, exposure visits, and demonstrations assured quick adoption of sustainable water management and agricultural practices.
- Well-trained and strongly-motivated staff of the Government and NGOs working closely with farmers is necessary for the successful implementation of PGM.

APWELL to APFAMGS

On the recommendation of the Mid-Term Review Mission†, the APWELL gradually initiated a number of pilot activities related to water conservation, including WSD in two villages, PHM in all clusters with more than 10 successful groundwater irrigation systems, an experiment with people-controlled groundwater system in upper Gundlakamma sub-basin in Prakasam District; artificial recharge measures in two watersheds (with technical inputs from the National Geographical Research Institute (NGRI)), and introduction of drip and sprinkler irrigation, and eco-farming through application of low-cost, bio-fertilisers and bio-pesticides. The APWELL Project also conducted water quality testing in fluoride-endemic areas. During the final year of the APWELL Project, it was decided that the Indo-Dutch development assistance agreements were not to be extended to new projects. Instead, the Dutch Government approved a far smaller capacity building initiative to support farmer-managed groundwater systems for implementation through a network of NGOs in the seven APWELL districts. This was called the APFAMGS, for which funding was provided directly by the Royal Netherlands Embassy (RNE) till June 2004, after which it was transferred to the FAO.

The APFAMGS Project was implemented in the same seven districts (Map 5.1) as that of the APWELL, covering 650 habitations in 66 HUNs. It works in partnership with groundwater-dependent farmers, and empowers farmers with the knowledge and skills to monitor the groundwater system and take up appropriate interventions towards its management. The APFAMGS Project adopted a sub-basin approach for selecting habitations, unlike the APWELL which selected villages with exploitable surplus of groundwater. Thus, the approach to groundwater management shifted from water sharing to water management. Moreover, its infrastructure and incentive-centred approach has transformed it into a scientific knowledge-intensive approach.

---

† AP Groundwater Borewell Irrigation Schemes (APWELL): Mid-Term Review Mission Report, Netherlands Economic Institute, 1997.

Map 5.1

Location Map of the Operational Area under the APWELL/APFAMGS Project

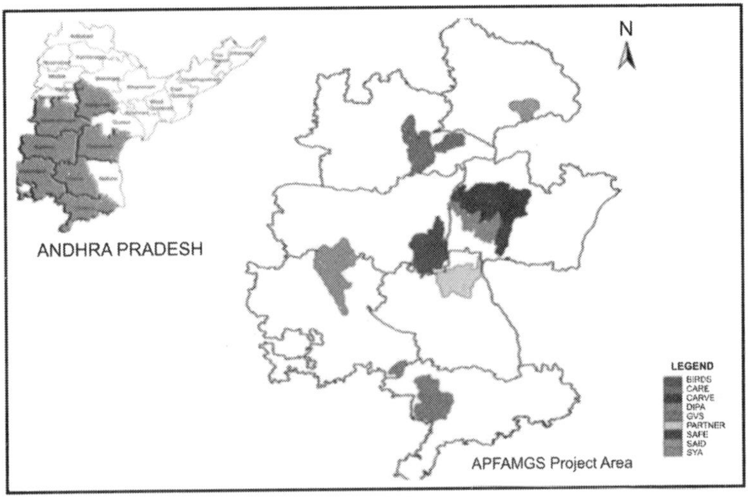

The philosophy of the APFAMGS Project is: 'farmers' understanding of groundwater dynamics makes the difference'. This is achieved through the process of enabling primary stakeholders to get involved in PHM for sustainable use of groundwater resources using hydrological boundaries as an operational unit. The APFAMGS Project is implemented through a network of Community-Based Organisations (CBOs) including nine field level partner NGOs and two international resource agencies. The main objective of the project is to 'equip groundwater farmer users with the necessary data, skills and knowledge to manage groundwater resources available to them in a sustainable manner, mainly through managing and monitoring their own demand'. The basic premise is that self-generated scientific data and knowledge will enable farmers to make appropriate farming choices using groundwater. The farming communities make informed decisions using hydrological data developed on the Geological Information System (GIS) platform. Elaborate institutional arrangements with equal representation of men and women were made to implement the programme.

Activities

The main activities taken up include:

- Awareness generation on the emerging groundwater crisis, and treating groundwater as a 'common good' at the habitation and hydrologic unit level.
- Demystifying the science of hydrology through participatory learning, practising and establishing a new relationship between farmers and groundwater.[†††]
- Participatory planning and sharing of information through CWB workshops for evolving common strategies that limit damage to the groundwater system without sacrificing individual interest.

Additional steps include:

- improving crop water efficiency,
- reducing chemical pollution,
- introducing groundwater governance transcending individual holdings and habitations without being coercive through voluntary choices such as reduced pumping, preventing construction of new wells, crop diversification, reduced application of chemical fertilisers/pesticides, etc.

Approach

A comprehensive institutional structure integrating technical and social components was established. At the village level a GMC is the key institution of the farmers – both men and women. A network of GMCs is formed at the hydrological unit level, viz. the HUN. These two are critical for providing a 'demonstration effect' of the learnings from the project to the larger community of farmers beyond the project area. The HUNs have a legal status, allowing them to receive funds as well as carry out business activities. Making the farmers water literate is the core of the approach. The first step in this direction is to enhance the farmer's capacities to collect and analyse data on their own.

††† However, it is not easy to demystify science using the approach of participatory learning.

Photo 5.1

GMC Meeting

Capacity building and training activities are part of project components. Formal and informal techniques such as technical training related to recording rainfall, measuring draft from observation wells, cultural shows, practical training, exposure visits, exchange visits, and workshops are included. These capacities are used in the PHM exercise. In PHM, the farmers volunteer to monitor water levels from 2,026 observation wells (one well for every sq km) every fortnight. Daily rainfall measurement is collected from rain gauge stations from 190 rain gauge stations established for every 5 sq km in the project area. The collected information is shared with the farmers for taking farming decisions. Discharge measurements are also carried out to understand the pumping capacity in 700 monitoring observation wells. This is accomplished by measuring the time taken to fill a drum of known capacity. Based on these measurements, the farmers have a good understanding of the pumping capacity of the wells, well performance, water requirement for different crops, and the ways and means to increase water use efficiency.

Crop Water Budgeting

The success of demystifying science is reflected in the CWB, which helps farmers to collectively make land use plans,

Table 5.3

A Sample of Groundwater Balance Estimates for a Few HUNs in 2008–2009

(in cubic metres)

HU Name	*No. of Habitations*	*Kharif Recharge*	*Kharif Draft*	*Kharif Balance (+ or –)*	*Rabi Recharge*	*Rabi Draft*	*Rabi Balance (+ or –)*
Chinneru	18	19,23,040	64,08,000	49,70,881	1,99,22,151	1,31,31,920	13,25,550
Rallavagu	15	17,85,521	42,55,000	7,60,671	1,21,10,183	53,49,920	41,50,503
Thundlavagu	7	14,86,319	45,24,000	25,65,112	1,16,28,227	71,30,900	9,59,832
Peddavagu	5	6,46,770	12,40,000	1,70,896	18,73,654	30,15,560	–17,62,674
Lothuvagu	1	3,42,844	5,82,000	46,692	6,96,252	2,91,400	1,61,869
Chandravagu	4	5,07,897	10,20,000	1,33,397	12,19,209	24,15,680	–17,27,864
Buchammakonetivanka	1	2,44,757	3,60,000	80,257	5,41,018	15,36,500	–11,22,631
Konetivanka	3	2,98,753	10,50,000	15,36,003	42,31,643	36,71,200	–4,94,390
Bavanasi	12	21,36,940	30,24,300	49,68,380	1,52,24,395	1,14,32,080	19,59,941
Yerravanka	4	6,06,641	18,00,000	2,71,304	32,39,133	54,76,720	–34,78,769
Peddavanka	4	3,44,320	26,19,000	23,11,120	95,31,133	66,31,520	9,772

Source: APFAMGS Project report, http://www.apfamgs.org.

depending upon water availability. The CWB is taken up at the village level before the starting of each season and aggregated at the HUN level. Using rainfall data and assumed run-off coefficient (10%), the contribution of rainfall to groundwater recharge is estimated. The net availability of groundwater is estimated by adding or deducting the previous season's balance (Table 5.3). There may either be positive or negative water balance in each season, depending on the recharge and draft. Based on the crop's water requirements and the net available groundwater, crop areas are decided in a collective manner. By following local measures, the volunteers explain the area under each crop with the available groundwater. They estimate the area that can be devoted to paddy if the entire water is used for paddy crop or other crops, or a combination of different crops. The estimates show that in 59 of the 63 HUNs, groundwater balance is deficit. The CWB also identified over-exploited aquifers, and water-harvesting measures such as injection wells were taken up in these aquifers. In some areas, abandoned open wells were also used to trap the flood flows and transfer them to the aquifers. Though there is no coercive mechanism to force the farmers to adopt collective decisions, a survey was conducted after every season on the extent to which collective decisions were made and discussed in the GMC. The data on actual cropping pattern is used to arrive at the actual draft; however, there is always a difference between estimated and actual draft. Though individual farmers' decisions are respected, the GMCs and HUNs are able to act as pressure groups to advocate change in cropping patterns, use of sustainable agricultural practices, and water saving technologies in some places.

Achievements

The achievements are drawn from the self-assessment reports of the APFAMGS, and independent evaluation reports of the World Bank (nd), FAO (2008) and AFPRO (2006), coupled with our field experience. All the physical achievements reported

(Table 5.4) by the end of 2007 are endorsed in the evaluation studies. The figures are quite impressive as most of the HUs (559 out of 650) have created the hydrological database and are managing (636 Community-Based Institutions (CBIs)) their groundwater. In fact, the data generated in the property of the GMC is being sold to outside agencies for the purpose of research. More than 4,000 farmers are trained to read maps and more than 10,000 farmers can handle hydrological equipment. It is assessed that some of the achievements have surpassed the targets (FAO, 2008).

During the field visits, we have observed the farmers presenting crop water budget estimates and taking the water table measurements. However, farmers are yet to be trained on using the GIS. Three hundred Farmer Water Schools (FWS)[3] have been established to train the farmers to equip them with technical and non-technical aspects of groundwater management. Hydro-Ecosystem Analysis (HESA), which is a decision-making tool for groundwater management, is being adopted and supported by recharge and discharge factors. Crop plans and management of groundwater is based on this analysis and observations.

This is the same sequence used for Agro-Ecosystem Analysis in the classical Farmer Field Schools (FFS) approach (FAO, 2008). This focus of FWS is on the active and common farmers who can apply them directly on farm and also share them with a lager audience. The FWS has sucessfully created the first batch of over 10,000 farmers who have already emerged as trainers to other farmers both under the project programmes as well as for the government-run FFS. Such a training and adaptation has demystified hydrology, which is a hidden source, and helped farmers in understanding the resource

3. Under the FWS, 10,000 farmers meet once in every 15 days through 300 water schools to understand groundwater changes in the respective area for the entire hydrological season. Based on their understanding, farmers adopt suitable modification in their agricultural practices that can lead to significant reductions in groundwater use.

available and dynamics. Sharing of informatin across HUs resulted in evolving common strategies, limiting the depletion of groundwater table.

Photo 5.2

Farmer Field School

Table 5.4

Physical Achievements of APFAMGS Programme (2007)

Indicator	*Achievement*
Number of farmers capable of reading maps	4,322
Number of farmers capable of handling hydrological equipment	10,076
Number of farmers updating Hydrological Monitoring Records (HMRs)	3,052
Number of GMCs using GIS	0
Number of GMCs having hydrological database	559
Number of GMCs sharing hydrological database	559
Number of farmers adopting alternative agricultural practices/inputs	14,281
Types of alternative agricultural practices promoted	80

Cont'd...

...Cont'd

Number of CBIs involved in groundwater management	636
Number of women on the committees of CBIs	2,060
Number of women farmer volunteers	1,175
Number of GMCs operating Crop Water Kiosks (CWK)	9
Number of GMCs advising farmers on crop choices based on CWB	559
Number of GMCs promoting alternative agriculture	559

Source: APFAMGS Project report, http://www.apfamgs.org.

Some of the important achievements include reduction in groundwater pumping in a number of HUs. In 14 of the 63 HUs groundwater pumping has been reduced significantly, while in 9 other HUs the reduction was moderate, though it is dependent on the reliability of groundwater balance estimates. Overall, despite the reduction in pumping in number of HUs, the reduction is not significant enough to have a drainage basin-level impact. Reduced water pumping has a direct bearing on area under paddy, as paddy is water intensive and the most preferred crop. In all, except in four HUs, the area under paddy

Photo 5.3

GMC Women Members Learning Monitoring of Water Levels

cultivation has come down, ranging from a few acres to several hundred acres. The farmers' experience showed that they incur crop losses whenever they do not follow the collective advice due to water scarcity. Crop diversification has taken place in favour of pulses, oil seeds, fruits, vegetables, flowers, etc. Farmers try to offset the losses due to reduction in paddy by growing other high value crops. The risk associated with commercial crops such as mono-culture, reduced area under food crops, and loss in soil fertility, is also being addressed simultaneously. Water saving devices such as Sprinkler and Drip Irrigation were introduced for crops such as groundnut, sunflower, bengalgram, chillies and horticultural crops. It is estimated that groundwater pumping was reduced by more than 8 per cent (equivalent to 5 mcm per year) over the project area due to water saving techniques.

The experience of APFAMGS proves that a comprehensive approach could benefit the farming communities, though in a limited way at present.

Impact of APFAMGS

Several impacts, on expected lines, are reported by the Project. Though some of the claims require technical verification, these impacts are:

- Empowerment of the community to collect, analyse and use data and knowledge related to water;
- Change in perception of groundwater from private property to that of a common good;
- Shift from cultivation of irrigated water-intensive crops to less water-intensive Irrigable Dry (ID) crops;
- Reduced losses from irrigated crops and increased profits from rain-fed or less water-intensive cash crops;
- Reduced groundwater draft;
- Increased groundwater recharge;
- Reduced use of chemical inputs;
- Increased use of organic methods of farming; and
- Reduced migration.

Social Regulations in Water Management at the Community Level (SRWM)

An action research project called 'Social Regulations in Water Management at the Community Level' (SRWM) was initiated in 2004 in three villages in AP by the Centre for World Solidarity (CWS), in partnership with local grass-root NGOs. Another village in Warangal was added during 2007. The project aims to promote local regulation and management of groundwater resources with equitable access to all families in the communities. The project is expected to develop models to equip the community with drought mitigation preparedness strategies through better water management and regulations at the community level; and to support Community-Based Organisations (CBOs) and *Panchayat Raj* Institutions (PRIs) in prioritising the needs of the community for drinking water, irrigation, and other uses, based on the principles of equity. Specific objectives of the project include:

- To develop the capacity of the community and NGOs on CWB, water supply and demand, and water balance assessments.
- To strengthen the role of PRIs and water communities, to decentralise decision-making, and creating the authority to enforce the rules, regulations and norms.
- To regulate water demand to ensure that everyone has access to at least the basic minimum of water for drinking and household purposes.
- To ensure regulatory mechanisms in irrigation practices that fit/relate with organisational structures such as Natural Resources Management (NRM) committees of *Gram Panchayats*, Watershed Committees (WSC), etc. and develop appropriate linkages to other NRM sectors.
- To crystallise facilitating mechanisms for social regulation of water resources and advocate the Government for wider replication and policy change.

The project is being implemented in four villages from three districts covering 715 households at an estimated cost of about ₹ 2.5 million per year over three years from AEI,

Luxembourg. The four project villages include Madirepalli and CR Pally in Anantapur District, Mylaram in Medak District and Enabavi in Warangal District. In all the four villages, rain-fed agriculture is the norm, but groundwater is an important contributor to irrigation for 6 to 42 per cent of the land. Groundwater provides the much needed life-saving irrigation during prolonged dry spells. Since 2009, as many as 15 more villages, which are spread in Anantapur, Chittoor and Nellore districts, were added in the project.

Prior to the 1990s, open wells with electrical centrifugal pumps were used to extract groundwater in the programme villages. Farmers started drilling borewells during early 1990s – the number of borewells grew rapidly in these villages over the last 15 years – and the shallow open wells gradually dried up due to declining groundwater levels. Due to indiscriminate drilling of borewells and unscientific groundwater exploration, many borewells failed either at the time of drilling or during later years. Furthermore, drilling borewells as deep as 300 ft at a closer spacing resulted in the drying up of the shallow, open dugwells, and shallow borewells due to well interference. Competition between neighbouring farmers often leads them to drill borewells as close as two metres apart. For instance, in Madirepalli Village, three neighbouring farmers dug 13 borewells in an area of 0.5 acres over a period of four years in competition to tap groundwater.

This phenomenon resulted in huge loss of investments to farmers and seriously affected the livelihoods of farmers dependent on irrigation. The project realised that there is need for changing the mindset of the farmers from 'competition' to 'cooperation' and to increase the 'water literacy' among the farmers for efficient use of water.

The project interventions began with a participatory assessment of water resources in the project villages. Participatory Rural Appraisal (PRA) methods were used to map the resource status and the existing water utilisation pattern for different purposes, such as drinking, domestic, and for irrigation. Growth of groundwater-based irrigation and trends in the groundwater levels over a period of time were thoroughly discussed and analysed in community level meetings, wherein women and men from all households participated. A series of such meetings and interactions helped to arrive at the crux of the issues, i.e. frequent failure of borewells and increasing debts of farmers due to investment in new borewells.

Photo 5.4

Assessing the Village Level Resources

Map 5.2

Location Map of SRWM Project Sample Villages

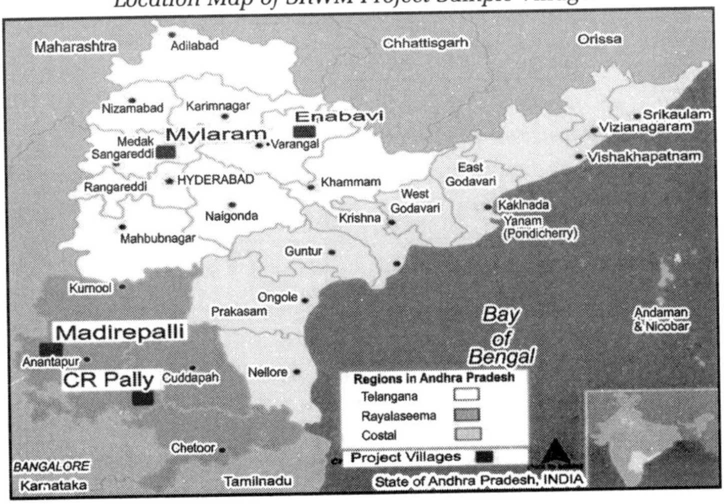

A number of training programs, exposure visits and awareness-raising meetings were organised by the grass-root partner NGOs supported by CWS in the project villages. Further public awareness and education was carried out through posters, pamphlets and wall writings. PHM of rainfall and groundwater levels in selected borewells was done regularly and shared and discussed at village meetings in order to increase the understanding of farmers on the behaviour of groundwater in relation to rainfall.

Photo 5.5

Officials Interacting with GMC

Photo 5.6

Exposure Visit to Water Efficient Paddy Cultivation

Photo 5.7
Measuring Water Level

Photo 5.8
Displaying Water Level Data on the Village Notice Board

A volunteer from the community measured rainfall from a simple manual rain gauge station installed in the villages and recorded the static water levels in 10 sample borewells using an electronic water level indicator. This data was displayed on a village notice board and updated periodically.

The first three years (from a total of seven years) of intensive grass-root work and facilitation has resulted in the community realising the ill-effects of indiscriminate drilling of borewells and use of groundwater. The community evolved and agreed on the following 'social regulations' and interventions in the village:

- No new borewells to be drilled in the village;
- Equitable access to groundwater for all the families through well sharing;
- Increasing the groundwater resources by conservation and recharge; and
- Efficient use of irrigation water through demand-side management.

Small groups of farmers were formed in all the project villages including a borewell owner and a set of about two to three neighbouring farmers, who did not own borewells. The borewell owners were motivated to share water by explaining that drilling new wells in the vicinity of their wells may render their's dry due to competitive extraction. Instead, sharing a portion of water from his well helps his neighbours, while securing his access to water and thus livelihood. Sharing water with neighbours is a 'win-win' situation, benefiting both the borewell owners and water receivers.

Sharing the Resource at the Village Level

Sharing of groundwater resource by well owners with other farmers is the prominent feature of the PGM in villages. The practice is of significance in an over-exploited area like Anantapur. For instance, Madirepalli Village of Singanamala Mandal has basically granitic terrain. The dug wells are very deep – sometimes more than 35 feet – and dry. The farmers in the village are practising groundwater management by sharing and conserving the resource through micro-irrigation and its augmentation through construction of recharge structures. Before the interventions, the farmers were given training on PHM, and are provided with the required equipment. However, there was slackness in practising it.

The villagers express gratefulness to the local NGO (RIDS) which created enough awareness among them about the futility of drilling new bores and the advantages of sharing water. Most of the farmers reported that earlier, they drilled numerous borewells of various depths. Most of these were drilled without scientific investigation, and huge money was invested in the hope of getting enough water. They believe that the efforts of the NGOs would ultimately pave the way for changing the groundwater scenario in their village. This change of practices occurred over a period of time, and also with the cost attached to the lessons learnt. Between 2004 (before the intervention) and 2010 (after the intervention), there was significant change in the attitude of the farmers, with substantial physical gains (Table 5.5). For instance, two of the open wells and 16 of the borewells were revived after the intervention. The area under irrigation also increased substantially, i.e. 31 per cent in the case of *Kharif* and 158 per cent in the case of *Rabi* crops. This was possible mainly due to water sharing, reduction in the cultivation of water-intensive crops (paddy), and increase in area under micro-irrigation. Groundwater recharge has been enhanced through renovation of recharge structures such as percolation ponds, check dams, etc. A notable achievement is that 78 farmers are sharing water with well owners and getting critical irrigation for their irrigated dry crops. The area under *Kharif* paddy has declined by 31 per cent, while *Rabi* paddy is totally stopped; and direct irrigation is used only in the case of paddy.

Table 5.5

Impact of SRWM Project in Madirepalli Village

Details	*Before (2003–04)*	*After (2010–11)*	*Change/ Impact*
Area under cultivation (acres)	767.5	767.5	–
No. of functional open wells	2 (59)	4 (59)	Increased
No. of functional borewells	53 (75)	69 (79)	Increased
Area irrigated	Kharif: 213 Rabi: 127	Kharif: 280 Rabi: 328	Increased by 31% in Kharif and 158% in Rabi
Number of observation borewells	0	10	Increased
Number of sharing groups formed	01	69	Increased
No. of farmers sharing water	08	78	Increased
Area under paddy (acres)	Kharif: 74 Rabi: 73	Kharif: 51 Rabi: 0	Kharif: –31% Rabi: –100%
Area under direct irrigation (acres)	314	51	–84
Area under micro-irrigation (acres)	26	557	Increased

Cont'd...

...Cont'd

Construction of recharge structures	0	28 (percolation tanks, check dams and recharge pits)	Increased recharge of wells
Cropping pattern	Paddy for regular consumption through very low-yielding BWs	Switched over to ID crops like groundnut, green chilli, sunflower, etc.	Better financial returns, conserved resource

Note: Figures in brackets are the total number of wells.

Source: Rural Integrated Development Society (RIDS), 2011.

Earlier, the villagers drilled a large number of bores individually in their lands in the hope of having an irrigation source. For instance, a farmer, K. *Subbanna*, drilled 26 bores but only one was successful and is functioning to this day. Most of the bores went dry or were low yielding. The farmers earlier resorted to growing paddy and other water-intensive commercial crops but subsequently switched over to Irrigable Dry (ID) crops such as sunflower, groundnut, etc. After the intervention, the farmers decided not to drill new borewells and share the water from the successful bores with other farmers. Water is being shared between brothers, among farmers irrespective of caste and between small and big farmers in the village. The villagers constructed 28 recharge structures and helped to augment the yields of the successful bores. The number of farmers sharing bores increased from eight in 2004 to 78 in 2010–11. The recharge structures constructed during the past 2–3 years reportedly revived/rejuvenated some of the defunct borewells and are presently irrigating about one to three acres per well. Earlier, farmers used the flood irrigation method, but now they are adopting micro-irrigation methods such as drip and sprinkler irrigation for the ID crops.

Key Achievements of the Project in Madirepalli Village

- Gradual change in thinking among the community, recognising groundwater as a scarce resource and CPR;
- Enhanced resource availability through rejuvenating and taking up new water harvesting activities;

- Created drinking water access to fulfil the entire community's and cattle's needs, however, these impacts are local and do not take the scale impacts (Syme et al., 2011);
- All 69 individually-owned irrigation borewells came under the water-sharing system providing water access to 78 new farmers;
- About 268 acres of rain-fed lands brought under protected irrigation by sharing water from borewells using micro-irrigation systems, this corresponds to 44 per cent of the total well irrigated area in the village during 2010–11;
- Relative extraction of groundwater reduced from 125 to 80 per cent of the annual available groundwater from the year 2004–2005 to 2010–2011; and
- Farmers changed from water-intensive crops to water-saving crops.

Though there was a deeper crisis in agriculture in Madirepalli Village due to higher groundwater dependency, the existence of traditional regulatory practices in *Gonchi* seepage channels and motivated village leadership contributed to better results of the project in Madirepalli compared to the other project villages. While Madirepalli was successful in expanding the water-sharing system to many borewells, there was a significant change in cropping pattern (from water-intensive crops to water-saving crops) in Mylaram Village. The project was also successful in building a community level institution, called Water Resources Committee, in CR Pally, which took up the agenda of groundwater management and regulation. From 2009, electricity efficiency measures such as formation of Distribution Transformer (DT) level farmers' groups; installing capacitors on all pump-sets; and regularising unauthorised electric connections to agricultural pump-sets helped to reduce low voltages at pump-sets; and contributed to reduction in motor burnouts in project villages.

Andhra Pradesh Drought Adaption Initiative (APDAI) Project

The APDAI pilot project is being implemented in two phases due to different modes of financing. Phase I of the pilot programme (April 2006–June 2007), financed by a World Bank-executed trust fund, initiated activities in six villages in three Mandals of

Mahbubnagar District. Phase II of the pilot implementation was started in November 2007 and the project was expanded into an additional nine villages in Mahbubnagar District and initiated activities in 10 new villages in Anantapur District. In addition, there is an option to pursue pilot initiatives outside the 10 selected villages in Anantapur. The implementation of the APDAI Phase II is being supported by the Japan Policy and Human Resources Development (JPHRD), the Climate Change Initiative Grant (CCIG), and the World Bank. The pilot activities are implemented by the Society for Elimination of Rural Poverty (SERP) in collaboration with District Collectors in the pilot districts, and under the oversight of the Principal Secretary, Department of Rural Development (DoRD) through the Office of the Commissioner, Rural Development (CRD).

The drought adaptation pilot is rooted in the strength of the CBOs and is implemented by the federation of women SHGs (*Mandal Mahila Samakhyas* – MMSs) in convergence with various Government departments. The pilot initiative relies on pooling existing experience and expertise of NGOs, research institutes, and CBOs into a consortium of supporting agencies led by WASSAN to facilitate the action research on the ground. As part of APDAI initiatives, a new approach was introduced to secure rain-fed crops through sharing groundwater for critical irrigation, and involving communities for management by developing social regulations. WASSAN is the lead technical agency for this pilot.

The APDAI approach of community groundwater management aimed at two shifts:

1. From individual farmer approach to area-based approach for irrigation; and
2. From groundwater as private property to groundwater as common property.

It aimed at building a case for enabling policy support and investments on critical/protective irrigation and water sharing, focusing on rain-fed farmers. The envisaged model included an approach to facilitate a common understanding between owners and non-owners of borewells, to share the groundwater. It also provided incentives for sharing and initiating social regulation for controlling the competitive digging of borewells. Further, farmers were supported with pipeline networks for transportation of water to rain-fed farms and linkage with micro-irrigation systems that contribute to maximise the groundwater use efficiency. The pilot was taken up during 2006–07, initially in Chellapur Village of Mahbubnagar District. Later it was extended to eight villages in Mahbubnagar, Anantapur and Ranga Reddy districts. The main objectives of the initiative include:

- Stopping the competitive digging of borewells;
- Providing access to the groundwater for rain-fed crops for protective/critical irrigation, which improves their productivity;
- Reducing water loss by adopting effective irrigation systems and methods;
- Reducing the cultivation of water-intensive crops (paddy) under borewells and motivating the farmers for alternative crops to improve water productivity;
- Enabling village level institutions for Community Managed Groundwater Regulations, including monitoring of groundwater level and borewell yields;
- Improving the groundwater recharge, in the long run through convergence; and
- Ensuring food and fodder security for household needs.

Implementation Process

The process is initiated with participatory analysis of agriculture under rain-fed conditions and the need for protective irrigation in order to make crop production viable. These exercises are usually carried out with the entire village and also with farmers in small groups. Area-based approach involves organising farmers under Common Interest Groups (CIGs) for a rain-fed patch. In each patch, well owners are convinced to share their water with the surrounding farmers. Once consensus is reached on water regulations and sharing the cost of pipeline installation, an agreement on groundwater regulation is signed by all the farmers in the patch in the presence of a *Tahsildar* on a bond paper for ₹ 100.

As per the agreement, all the borewells will be pooled through a common pipeline network and water will be shared among all, irrespective of ownership. No new borewells will be dug for at least the next 10 years. The cropping pattern will be decided on the basis of crop plans linked to the availability of water in agreement with members of the CIG while giving priority to food and fodder crops and a reduction in the area under paddy. One borewell a day will be rested on rotation, thus reducing water pumping by about 20 per cent. While water is shared to protect the *Kharif* crop of non-owners, the acreage of borewell-owning farmers is ensured; and a general fund is created for the maintenance of pipeline, repairs, etc. within the CIG.

The water from the borewells of the farmers willing to share is interconnected to one main pipeline, which is distributed to the identified rain-fed patch of land. This involves:

- Preparation of a pipeline network plan:
    - Identification of borewell points;
    - Levelling survey using hydrometer for main pipeline, sub-lines and outlets;
    - Plot-wise area measurement and location of pipeline on the field; and
    - Preparation of a detailed map showing the individual plot and total patch area boundaries and the pipeline network.
- Preparation of pipeline design and detailed estimation of the pipeline network.
- Work execution:
    - Calling quotations from reputed PVC companies;
    - Placing a work order to the shortlisted company; and
    - Execution of work by CIGs and *Grama Samakhyas*, with the support of WASSAN.
- Evolving crop planning and regulation mechanisms:
    - Restriction of water-intensive crop area for each farmer owning borewells;
    - Forming outlet-wise groups for water use;

- Appointment of one or two persons for water distribution;
- Formation of water regulation committee at village level;
- Outlet-wise scheduling of water distribution at the time of critical irrigation; and
- Developing a common fund for pipeline maintenance by collecting water charges from the water users of the pipeline network.

As there was no threat of new borewells in the vicinity that may lead to the drying of their own borewell, the farmers agree to pool their borewells and share the water. This would avoid competitive borewell digging, unnecessary investments, and loss of capital. The borewell owner is assured of earlier cropped area but with less water-requiring crops. The water thus saved will provide critical irrigation to a rain-fed patch, which includes lands of both borewell owners as well as others. If any one of the borewells fails, there is a back up arrangement as they are pooled. There was also motivation in terms of getting access to micro-irrigation system (sprinklers and drips) at subsidy, through linkage with the Andhra Pradesh Micro-Irrigation Project (APMIP). This was intended to increase the groundwater use efficiency. The APDAI has also extended up to 90 per cent support for pipeline network required for water sharing.

Impact of the Pilot

- Able to provide protective irrigation for selected rain-fed patches in the pilot villages;
- Ensured timely sowing, especially during delayed monsoons (because of assured water supply);
- Increase in cropped area under the pooled borewells;
- Incremental returns on crop yield;
- On an average, about 25 to 30 per cent of the pumping hours were saved through resting of wells, resulting in saving both the groundwater and power consumption;
- Micro-irrigation system and pipelines have reduced the labour time for irrigating the crop (seven hours to one hour);

- It also increased water use efficiency; and
- Arresting competitive digging of borewells.

Experience in Gorantlavaripalle

Gorantlavaripalle in Nallacheruvu Mandal has 113 families and the total cultivated area is 270 acres. There are 26 borewells, 60 families having irrigated lands, and 40 families without any source of irrigation. Normally, paddy is cultivated under borewell irrigation. Farmers with borewells also have drylands, which are at a distance of more than one kilometre from the borewells, while farmers without any water source have lands nearer to the borewells.

To facilitate groundwater sharing, all the farmers in the village were organised into five groups, based on the contiguity of the land. Thus, the village area was divided into five blocks. Each block consists of farmers with and without wells. All the groups passed a resolution agreeing to network the 26 borewells through a single pipeline. The farmers in their groups identified those with water and without water and came to an agreement regarding who would share water with whom. A committee was formed with two representatives from each group (one with water source and the other without water source) and a representative from the Village Organisation (VO). Based on the agreement, WASSAN facilitated the MMS to undertake the following surveys:

1. Ground levelling survey – This was completed in October, 2008;
2. Pumping water level – This was completed in November, 2008;
3. Static water level survey; and
4. Discharge measurement.

The last two surveys should be done once in a month. After the above surveys, a plan and a budget estimate were prepared for networking the borewells and distribution pipes. They also framed norms and regulations and signed the agreement, with the *Tahsildar* as a witness, on a ₹ 100 stamp paper. The plan and estimates were discussed in a meeting organised by the MMS. It was estimated that the cost of networking and distributing water to dryland crops would cost about ₹ 8 to 12 thousands per acre. Based on the discussions in the MMS, it was proposed that the cost should be shared in the following proportion: 25 per cent as initial farmer's share; 25 per cent as subsidy from the programme; and 50 per cent as loan to the farmer (either from bank or the VO).

However, farmers conveyed their inability to bear these costs due to frequent droughts in the recent years. A meeting was organised in the village in June, 2009 with the farmers, staff of MMS, WASSAN, and the Project Directors of District Water Management Agency (DWMA) and District Rural Development Agency (DRDA). It was agreed that farmers would first pay a membership fee of ₹ 1,000 per acre and another ₹ 1,000 per acre after obtaining the crop yields. The canal digging work was initiated in July, 2009, and was completed in December, 2009.

All the farmers under the borewell network prepared a crop plan for *Kharif*, 2010. They agreed to decrease the area under paddy and other water-intensive crops, such as sugarcane, under the borewells. For efficient functioning of the networking system they agreed to appoint one *Neerugatti* (waterman) with the following responsibilities:

1. Encourage every farmer to take up crop cultivation;
2. Inform the farmers about their turns for water sharing;
3. Collect annual fees of ₹ 1,000 per acre after the yields are obtained;
4. Out of the amounts collected the share of the waterman is 20 per cent and the remaining 80 per cent is allocated for repairs and maintenance – this amount is deposited into the bank account of the committee; and
5. Bring at once to the notice of the committee members regarding leakages and repairs.

In the next chapter we assess the impact of the CBGM institutions at the household level. This is based on the quantitative and qualitative data collected from the sample households from the sample village.

6

Approach and Profile of Study Sites

Approach

For the purpose of comparative institutional assessment, three villages were selected, where the community groundwater management practices have been adopted under different NGOs. Besides, one village where no such management practices were adapted was selected as the control village. The details of the sample villages are presented in Table 6.1. The sample villages consist of:

1. One village covered under the APFAMGS implemented by the NGO, Bharati Integrated Rural Development Society (BIRDS). APFAMGS is a continuation of an earlier programme known as APWELL Project supported by the Royal Netherlands Government. The APFAMGS was implemented with the support of Food and Agricultural Organisation (FAO);
2. One village, where the NGO, CWS, along with its local partner NGOs, has been implementing the programme named 'Social Regulations in Water Management' (SRWM);
3. One village, where participatory groundwater management is being promoted as part of the APDAI of the Department of Rural Development (DoRD). The NGO, WASSAN is the lead technical agency, and the initiative is being

Table 6.1

Details of the Sample Villages

Village	*Mandal*	*District*	*Groundwater Model/ Project*	*Implementing Agency (NGO)*	*Year of Initiation of Project*	*Stage of the Project**
Thaticherla	Komarolu	Prakasam	APFAMGS	DIPA (BIRDS)	2003–04	1st phase completed 2nd phase ongoing
Madirepalli	Singanamala	Anantapur	CWS/SRWM	RIDS (CWS)	2003–04	Ongoing
Gorantlavaripalle	Nallacheruvu	Anantapur	WASSAN/APDAI	WASSAN/MMS	2007–08	Ongoing
Rajupalem	Komarolu	Prakasam	Control Village	NA	NA	NA

NA: Not Applicable

* Refers to the status of the project during the field survey.

Note: At present all these Projects are completed.

implemented by the SERP, through the *Mandal Mahila Samakhyas* (MMS); and

4. One control village with substantial groundwater use, but not having any groundwater management institutions.

Of these three initiatives, the APFAMGS Project operates at a wider scale, covering 3,000 farmers in seven districts of AP, while the other two are working on an experimental basis on a small scale of a few villages. Though the APFAMGS initiative focuses on rain-fed and semi-arid regions, the socio-economic, agro-climatic and hydro-geological conditions vary widely across the locations and villages. An attempt was made to select a representative village from across the seven districts to identify the common elements in the institutional arrangement and the processes that may be common and relevant for comparison with other initiatives.

Profile of the Sample Villages

The sample villages vary in size (number of households) and socio-economic composition. The control village is the largest, with 374 households; while the smallest is Gorantlavaripalle, with 113 households (Table 6.2). The geographical area of the sample villages ranges between 300 and 1,900 hectares, and the average family size ranges from 3.8 to 4.4 (Table 6.3). Socially, two of the sample villages are dominated by the Other Caste (OC) households, while two of them have a higher proportion of Backward Caste (BC) households. Two of the sample villages have more than 25 per cent of the households belonging to the Scheduled Caste/Tribe (SC/ST) households. In terms of economic composition, most of the sample villages are dominated by marginal and small farmers (Table 6.2). Only Madirepalli Village has about 50 per cent of the households from medium and large farmers. These variations help in understanding the dynamics of CBGM in varying socio-economic contexts. The proportion of the sample from these villages ranges between 8 and 27 per cent. This is due to the size of the sample village, as the number of sample households chosen are 30 from each sample village.

Table 6.2

Socio-Economic Composition of the Households in the Sample Villages

District/ Farm Size	*Village/Social Categories/No. of HHs*			*Total HHs*	*Sample HHs*
	SC/ST	*BC*	*OC*		
	Prakasam: Thaticherla				
Landless	10	30	5	45	0
Marginal Farmers	38	100	3	141	19 (14)
Small Farmers	10	40	15	65	8 (12)
Medium Farmers	2	0	10	12	3 (25)
Large Farmers	0	0	2	2	1 (50)
Total	60	170	35	265	31 (12)
	Anantapur: Madirepalli				
Landless	2	6	1	9	0
Marginal Farmers	26	8	9	43	5 (14)
Small Farmers	3	4	30	37	8 (22)
Medium Farmers	0	25	30	55	12 (22)
Large Farmers	0	7	22	29	5 (17)
Total	31	50	92	173	31 (18)
	Anantapur: Gorantlavaripalle				
Landless	7	0	0	7	0
Marginal Farmers	19	20	5	44	8 (18)
Small Farmers	2	30	20	52	19 (37)
Medium Farmers	0	0	10	10	3 (30)
Large Farmers	0	0	0	0	0
Total	28	50	35	113	30 (27)
	Prakasam: Rajupalem				
Landless	28	0	4	32	0
Marginal Farmers	1	60	100	161	12 (8)
Small Farmers	1	10	150	161	13 (8)
Medium Farmers	0	0	10	10	3 (30)
Large Farmers	0	0	10	10	3 (30)
Total	30	70	274	374	31 (8)

Source: Field Survey (PRA/FGD Methods).

Note: Figures in the brackets indicate the per cent of sample farmer HHs taken for the study.

Access to Groundwater

Access to groundwater and irrigation is at the core of groundwater management. The extent and nature of access across the sample villages would highlight the differences in the functioning and performance of the PGM. All the sample villages depend on groundwater irrigation. The extent of irrigation ranges between 15 per cent in Gorantlavaripalle to 34 per cent in Madirepalli (Table 6.3). On the other hand, more than 70 per cent of the households in the three villages, where groundwater institutions are present, have access to wells, as against 29 per cent in the control village. Variations in the extent of irrigation (percentage of area under irrigation) could be due to the groundwater potential in the respective villages. However, the contrast in the access to wells in one form or the other, explains the role of groundwater institutions. For instance, though only 15 per cent of the households in Thaticherla own wells, 70 per cent of them have access to groundwater through water sharing and community wells. On the contrary Rajupalem (control village) has only 29 per cent of the households reporting access to well water for irrigation, despite 17 per cent of them owning wells. In the control village, only 12 per cent of the households share water with others as against 37 to 46 per cent of the households in the villages with groundwater institutions (Table 6.3).

There is a clear pattern in the access to groundwater across socio-economic groups of farmers. It is observed that the SC/ST farmers and marginal and small farmers seem to depend more on sharing water, while a large proportion of the OC farmers and large farmers have their own wells (Tables 6.4 and 6.5). The landholding pattern is more or less similar in all the sample villages, though we do not have specific information of the hydro-geology of the villages. These two factors are critical in influencing the access and quality of groundwater.

Table 6.3

Groundwater Access to Households in the Sample Villages

Particulars	*Thaticherla*	*Madirepalli*	*Gorant-lavaripalle*	*Rajupalem*
No. of Households	265	173	113	374
Average Household Size	4.4	4.2	4.3	3.8
Total Geographical Area (in ha)	1,903	307	1,064	1,676
Area under Irrigation (%)	32	34	15	22
% of HHs with Own Wells	15	43	26	17
% of HHs Sharing Wells	46	45	40	12
% of HHs depending on Community Wells	10	0	0	0
% of HHs with Access to Wells	71	88	87	29
Main Occupation	Cultivation	Cultivation	Cultivation	Cultivation

Source: Field Survey/Village Records.

Table 6.4

Details of Well Status of Groundwater Farmers Across Social Category and Farm Sizes in Sample Villages

District/Village/ Caste Category	*Well Status of Groundwater Farmers*			
	OW	*WS*	*CW*	*All*
	Prakasam: Thaticherla			
SC/ST	7 (1)	27 (4)	6 (1)	40 (6)
BC	26 (4)	79 (13)	15 (2)	120 (19)
OC	6 (2)	16 (3)	5 (1)	27 (6)
Total	39 (7)	122 (20)	26 (4)	187 (31)

...Cont'd

...Cont'd

	Anantapur: Madirepalli			
SC/ST	3 (1)	14 (4)	0	17 (5)
BC	18 (3)	26 (5)	0	44 (8)
OC	53 (11)	38 (8)	0	91 (19)
Total	74 (15)	78 (16)	0	152 (32)
	Anantapur: Gorantlavaripalle			
SC/ST	2 (1)	5 (2)	0	7 (3)
BC	15 (4)	27 (7)	0	42 (11)
OC	40 (12)	10 (4)	0	50 (16)
Total	57 (17)	42 (13)	0	99 (30)
	Prakasam: Rajupalem			
SC/ST	0 (0)	0 (0)	0 (0)	0
BC	7 (2)	9 (3)	0 (0)	16 (5)
OC	55 (16)	35 (10)	0 (0)	90 (26)
Total	62 (18)	44 (13)	0 (0)	106 (31)

Source: Field Survey (PRA/FGD Methods).

Note: 1. OW-Own Well; WS-Water Sharing; CW-Community Well.
2. Figures in the brackets indicate the No. of sample groundwater farmer HHs taken for the study.

Table 6.5

Distribution of the Sample HHs Across Farm Size and Well Ownership Status

Village	*Groundwater User Well Status*	*Economic Class*			*Overall*
		MF	*SF*	*LMF*	
Thaticherla	Owned	47	29	24	55
	Water Sharing	79	21	0	45
	Total	61	26	13	100
Madirepalli	Owned	7	40	53	48
	Water Sharing	31	13	56	52

Cont'd...

...Cont'd

	Total	19	26	55	100
Gorantlavaripalle	Owned	6	76	18	57
	Water Sharing	54	46	0	43
	Total	27	63	10	100
Rajupalem	Owned	22	44	33	58
	Water Sharing	62	38	0	42
	Total	39	42	19	100
Overall	Owned	21	48	31	54
	Water Sharing	55	29	16	46
	Total	37	39	24	100

Source: Field Survey.

Note: MF – Marginal Farmers; SF – Small Farmers; LMF – Large and Medium Farmers.

Methodology

Qualitative as well as quantitative research methods have been used for the study. Primarily, Focus Group Discussions (FGDs) and household questionnaires were used to elicit the required information. Besides, basic secondary data about the villages were collected from the village secretary, elders, and key informants. Field research was conducted during the months of February and March, 2011. The study team collected information and held discussions with key professionals involved in APWELL/APFAMGS, CWS, and APDAI/WASSAN projects for a broader understanding on the objectives and processes involved in the design and implementation of the respective initiatives. The study team also interacted with the officers and consultants of the State Irrigation and Command Area Development (I and CAD) and the Ground Water Department both at the state and district levels. During the field visits, the team had discussions with the staff of local NGOs implementing the respective programmes. Important issues covered include communication and awareness strategy, community participation, groundwater management by community, impact on cropping pattern and yields, etc.

For the purpose of quantitative household data collection, a detailed questionnaire was prepared, covering socio-economic, demographic, agriculture and groundwater management. In each village, about 30 households representing the socio-economic categories of the community were selected. The sample was divided into two groups, viz. well owners and those sharing wells or depending on community wells. The sample was divided in proportion to the actual number of well-owning and well-sharing households. At the end of the field visit, the gist of the information collected was shared with the villagers for the purpose of triangulation. Community wells are present only in one sample village (Thaticherla) under the APWELL/APFAMGS programme.

Results and Discussions

Impact assessment is carried out at two levels, viz. well owners and water-sharing farmers across farm sizes. Impacts are assessed not only between different types of institutions but also with and without institutions, i.e. using the control village. Three indicators, viz. access to irrigation, access to critical irrigation, and moving towards less water-intensive crops are assessed. Besides, awareness and perceptions of the farmers regarding the role and effectiveness of the institutions is also gauged.

Access to irrigation has gone up in all the sample villages, including the control village. It may be noted that the sample households include only those farmers having wells or those sharing water from well owners and hence, the proportion of area irrigated is on the higher side. The increase is the highest in the control village at 213 per cent (Table 6.6). The difference between the control and other sample villages is that the increase in area under irrigation is mainly through sharing of wells in the villages with institutional arrangements while in the control village, irrigation through own wells has gone up substantially. Across the size classes, increased access to irrigation is more among marginal and small farmers in three of the sample villages, including the control village (Table 6.7). On

Table 6.6

Changes in Percentage Area Under Well Irrigation by Well Status

Status	*Thaticherla (APFAMGS)*			*Madirepalli (SRWM)*			*Gorantlavaripalle (APDAI)*			*Rajupalem (Control)*		
	O	*WS*	*All*	*O*	*WS*	*All*	*O*	*WS*	*All*	*O*	*WS*	*All*
Before	56	30	49	49	8	31	77	18	59	32	0*	22
Present	76	83	78	60	63	62	92	64	83	66	77	69
% Change	36	176	59	22	688	100	19	255	41	138	–	213

Source: Field Survey.

Note: * Though there was the practice of sharing wells before 2004, there was no area covered as the groups became defunct, consequent to the drying up of wells. Hence, the changes are not entirely due to increased well-sharing activity; instead it is due to the revival of borewells under water sharing.

O– owned well, WS – water sharing, All – both owned well and water sharing together.

Table 6.7

Changes in Percentage Area Under Well Irrigation by Farm Size

Status	*Thaticherla (APFAMGS)*			*Madirepalli (SRWM)*			*Gorantlavaripalle (APDAI)*			*Rajupalem (Control)*		
	MF	*SF*	*LMF*	*MF*	*SF*	*LMF*	*MF*	*SF*	*LMF*	*MF*	*SF*	*LMF*
Before	50	62	25	25	39	30	8	66	80	17	14	41
Present	79	79	75	70	58	59	69	85	90	73	69	66
% Change	58	27	200	180	49	97	763	29	13	329	393	61

Source: Field Survey.

Note: MF – Marginal Farmers, SF – Small Farmers, LMF – Large and Medium Farmers.

Table 6.8

Changes in Access to Wells and Access to Water

Village	*Total No. of HHs (Population)*	*Period*	*Water-Sharing HHs*	*Area Under Paddy (acres)*	*Area Under Irrigation (acres)*	*Source of Irrigation*			
						Dugwells		*Borewells*	
						No.	*Area (acres)*	*No.*	*Area (acres)*
Thaticherla	265	B	45	132	168	24 (0)	0	30 (15)	38 (22)
	(1,155)	A	148	55	329	24 (0)	0	36 (31)	159 (48)
Madirepalli	173	B	8	180	254	59 (2)	4	75 (53)	200 (79)
	(725)	A	78	50	491	59 (4)	16	79 (69)	390 (79)
Gorantlavaripalle	113	B	10	128	140	34 (0)	0	82 (40)	90 (64)
	(487)	A	42	80	188	34 (0)	0	84 (46)	138 (73)
Rajupalem	374	B	25	150	199	9 (6)	14	40 (35)	95 (48)
	(1,414)	A	44	150	249	9 (4)	8	79 (62)	150 (60)

Source: Field Survey (PRA/FGD Methods) and Village Records.

Note: Figures in brackets are functional wells and percentage of area in the case of area.
B – Before, A – After.

the other hand, the large farmers gained more in the case of Thaticherla Village where APFAMGS is working. It may be noted that water sharing is neither new nor attributed to the groundwater institutions alone, and sharing has been practised in all the sample villages prior to the advent of these institutions. Even in the control village, well sharing has been practised, though on a limited scale between relatives, in the recent years. The role of institutions becomes clear in terms of other impacts such as reduction in individual wells, availability of critical irrigation and reduction in the cultivation of water-intensive crops.

The number of households sharing water has gone up in all the sample villages (Table 6.8), and the increase is substantially higher among the villages with groundwater institutions than in the control village. On the other hand, the number of wells almost doubled in the control village as against the moderate increase in the institutional villages. The number of functional wells has also gone up in all the sample villages. This could be due to the better rainfall conditions after 2004 when compared to severe drought conditions (three successive droughts) between 2001 and 2004. Most of the dugwells dried up during this period and a few of them revived after 2004. More importantly, investment in new wells is marginal in the sample villages where social regulation is in place (Madirepalli and Gorantlavaripalle), whereas in the case of APFAMGS village (Thaticherla), the number of borewells has gone up by 20 per cent, as there is no regulation.

Improved groundwater conditions in the sample villages under groundwater institutions are also evident from the availability of irrigation during critical periods. The number of farmers reporting availability of groundwater during critical periods has gone up in all the institutional villages, while the number has gone down in the control village (Table 6.9). However, this is limited to well owners in two of the villages. In the case of Gorantlavaripalle (APDAI), even the well-sharing farmers have reported that they have received critical irrigation. The marginal and small farmers are the main beneficiaries in terms of receiving critical irrigation in the institutional villages (Table 6.10), whereas in the control village, the proportion of marginal and small farmers receiving critical irrigation has come down. This indicates that groundwater institutions have improved the source sustainability and helped in protecting the crops to a large extent. This has been possible due to the reduction in the area under water-intensive crops (paddy) in two of the institutional villages (Table 6.11). However, the APFAMGS village, along with the control village, reported an increased area under paddy. The reduction in area under paddy in the institutional villages is more among large farmers, while the increase in area under paddy is more among marginal and small farmers in case of Thaticherla (Table 6.12). This is mostly compensated by groundnut crop. In all the sample villages, no crop area has declined substantially. The decline is more in the villages with social regulation. This reflects the improved access to critical irrigation. In the absence of any social regulation, the farmers do not seem to follow conservation methods, though they tend to reduce their risk of investing in new borewells as they are familiar with the groundwater situation due to the interventions of the APFAMGS.

Table 6.9

Availability of Irrigation During Critical Periods of Crop Growth by Well Status

(% of Farmers)

Availability/ Status	*Thaticherla (APFAMGS)*			*Madirepalli (SRWM)*			*Gorantlava-ripalle (APDAI)*			*Rajupalem (Control)*		
	O	WS	All	O	WS	All	O	WS	All	O	WS	All
Before	23	0	14	0	0	0	60	28	51	34	0	23
Present	36	0	22	10	0	50	77	67	74	20	0	13
% Change	60	–	60	–	–	–	29	140	46	–43	–	–43

Source: Field Survey.

Note: *O – owned well, WS* – water sharing, *All* – both owned well and water sharing together.

Table 6.10

Availability of Irrigation During Critical Periods of Crop Growth by Farm Size

(% of Farmers)

Availability/ Status	*Thaticherla (APFAMGS)*			*Madirepalli (SRWM)*			*Gorantlava-ripalle (APDAI)*			*Rajupalem (Control)*		
	MF	*SF*	*LMF*	*MF*	*SF*	*LMF*	*MF*	*SF*	*LMF*	*MF*	*SF*	*LMF*
Before	18	10	0	0	0	0	0	56	67	27	32	0
Present	27	20	0	17	75	50	50	79	67	5	16	21
% Change	50	100	0	–	–	–	–	41	0	–83	–50	–

Source: Field Survey.

Note: MF – Marginal Farmers, SF – Small Farmers, LMF – Large and Medium Farmers.

Table 6.11

Shifting Away from Paddy Crop by Well Status

(% area)

Crops/ Status	*Thaticherla (APFAMGS)*			*Madirepalli (SRWM)*			*Gorantlava-ripalle (APDAI)*			*Rajupalem (Control)*		
	O	*WS*	*All*	*O*	*WS*	*All*	*O*	*WS*	*All*	*O*	*WS*	*All*
Before												
No Crop	11	9	10	13	61	34	13	29	18	13	32	19
Paddy	11	4	9	63	4	38	23	4	17	14	0	10
Groundnut	7	4	6	24	35	29	29	46	34	11	0	8
After												
No Crop	9	9	9	5	2	4	3	0	2	14	12	13
Paddy	16	22	18	17	4	12	21	4	16	19	6	15
Groundnut	7	0	5	65	94	78	37	71	48	4	3	4
% Change												
No Crop	–17	0	–12	–63	–97	–89	–75	–100	–88	10	–64	–29
Paddy	50	400	100	–73	0	–69	–7	0	–7	36	0	55
Groundnut	0	–100	–20	173	171	172	28	54	39	–67	0	–56

Source: Field Survey.

Note: O – owned well, WS – water sharing, All – both owned well and water sharing together.

Table 6.12

Shifting Away from Paddy Crop by Farm Size

(% area)

Crops/ Status	*Thaticherla (APFAMGS)*			*Madirepalli (SRWM)*			*Gorantlavari-palle (APDAI)*			*Rajupalem (Control)*		
	MF	*SF*	*LMF*	*MF*	*SF*	*LMF*	*MF*	*SF*	*LMF*	*MF*	*SF*	*LMF*
Before												
No Crop	13	4	13	50	19	36	38	15	10	30	18	9
Paddy	8	13	6	15	55	36	0	16	40	3	12	13
Groundnut	13	0	0	35	26	28	62	31	20	3	6	16
After												
No Crop	11	13	0	5	10	0	0	3	0	13	14	13
Paddy	24	17	6	10	16	10	0	16	30	10	14	22
Groundnut	8	0	6	75	68	84	92	40	40	3	4	3
% Change												
No Crop	–20	200	–100	–90	–50	–100	–100	–80	–100	–56	–22	33
Paddy	200	33	0	–33	–71	–73	0	0	–25	200	17	75
Groundnut	–40	0	0	114	162	200	50	29	100	0	–33	–80

Source: Field Survey.

Note: MF – Marginal Farmers, SF – Small Farmers, LMF – Large and Medium Farmers.

Hence, the number of farmers growing paddy has increased in some cases, while in others, a decline was noticed. It is observed that a large number of farmers have started growing groundnut in Madirepalli.

The perceptions of the farmers in the institutional villages indicate high awareness about the institutions (Table 6.13). While the membership is limited to well owners in the case of APFAMGS villages, even the well-sharing farmers are members of the other two villages. As a result, institutional membership is quite low in the APFAMGS village. However, in all the villages, most of the sample farmers participate in the field or farmer schools participation rates range between 73 and 100 per cent among sample villages. On the other hand, participation in crop water budgeting is as low as 40 per cent in the APDAI village (Gorantlavaripalle). It was observed that all the farmers who participated in crop water budgeting exercise followed the recommendations in the social regulation villages while fewer farmers followed the recommendations in the APFAMGS village.

The main benefits perceived due to the institutions are awareness about groundwater, followed by crop methods, and groundwater irrigation methods (Table 6.13). Among the reasons for non-participation is the absence of tangible benefits followed by non-feasibility. While 70 per cent of the non-participating farmers felt that there are no tangible benefits in the APFAMGS and APDAI villages, only 35 per cent of the farmers perceived this reason in the case of SRWM village (Madirepalli). This perception is greater among the well-sharing farmers when compared to the well owners. Similarly, 81 per cent of the sample farmers in the APFAMGS village have endorsed the benefits from groundwater institutions, while 100 per cent agreed about the benefits in the other two villages. Lack of benefits is attributed to the reason that farmers do not follow the suggestions of the management committee, as the institutions play only an advisory role. However, the sample farmers in APFAMGS and APDAI villages perceive that the advisories are being followed or adopted.

Overall, the performance in terms of physical indicators and farmers' perceptions appears to be better in case of Madirepalli Village (SRWM) where social regulation is in place; while the performance of APFAMGS where there is no regulation seems to be poor. The APFAMGS initiative is the oldest among the three models. In fact, during the field work, the APFAMGS interventions were terminated, as the NGOs were waiting for the extension of the project. Hence, the poor performance of APFAMGS raises the issue of institutional sustainability (Reddy et al., 2011), and this is applicable even

Table 6.13

Farmers' Perceptions on Community-Based Groundwater Management (%)

Awareness on Groundwater Management Practices	*Details of Perceptions*	*Thaticherla (APFAMGS)*			*Madirepalli (SRWM)*			*Gorantlavaripalle (APDAI)*		
		O	*WS*	*All*	*O*	*WS*	*All*	*O*	*WS*	*All*
APFAMGS/SRWM/APDAI	Awareness	100	100	100	100	100	100	100	100	100
Membership	Yes	35	0	19	93	94	94	76	92	83
Participated in FFS	Yes	100	79	90	100	100	100	82	62	73
Benefits derived	Awareness about crops	100	71	87	100	100	100	82	77	80
	Groundwater methods	71	71	71	100	100	100	82	69	77
	Groundwater awareness	100	100	100	100	100	100	94	77	87
	All of the above	90	81	86	100	100	100	86	74	81
Reasons for not participating	No tangible benefit	59	86	71	33	38	35	59	85	70
	Not feasible	41	36	39	7	6	6	41	62	50
	Personal reasons	6	7	6	0	0	0	12	15	0

Cont'd...

...Cont'd

Participated in crop-water budgeting	Yes	100	43	74	100	100	100	65	8	40
Followed recommendations	Yes	82	29	58	100	100	100	65	8	40
Benefits from groundwater management	Yes	100	57	81	100	100	100	100	100	100
	Conduct of FSS/FWS/CWB	100	100	100	100	100	100	100	100	100
	Management of groundwater	88	86	87	100	100	100	82	85	83
	All of the above	96	95	96	100	100	100	94	95	94
Reasons for lack of benefits	Institutions play only advisory role	18	29	23	100	94	97	18	77	43
	Farmers did not follow GMC's suggestions	82	71	77	0	6	3	82	23	57

Source: Field Survey.

Note: O – owned well, WS – water sharing, All – both owned well and water sharing together.

for the other two initiatives. The difference between the other two initiatives is that the APDAI initiative is backed by the DoRD, while the SRWM is NGO-driven. The better performance of SRWM could be due to the intensive approach it has adopted in promoting water sharing – it has taken almost three years to organise the farmers and build awareness before initiating the well-sharing process. Besides, the SRWM worked with small groups of well-owning and well-sharing farmers, whereas the groups were bigger in the area-based approach followed by the APDAI.

7

Lessons for Up-Scaling

AP, arguably, has more experience in promoting community-based water management than any other Indian state. Even in the case of groundwater management, AP is the first state in the country to introduce community-based management way back in the 1990s. Unlike in the case of surface water, canals or tanks, there is no evidence of *in situ* institutional innovations in the case of groundwater. This is mainly due to the existing private (*de facto*) property rights on groundwater. Though these initiatives under study are still at a pilot stage, they can provide valuable insights for designing appropriate policies. However, the potential for up-scaling is linked to the specific hydro-geological and socio-economic settings and hence needs region-specific or flexible approach. Here we assess the strengths and weaknesses of the three institutional approaches and explore the possibilities for scaling up or policy lessons for bringing groundwater under community management.

The three models considered here have the common goal and objective of sustainable groundwater management. All the three institutions are led by NGOs with support from different agencies including the State Government. However, the approaches followed and the implementation modalities are different and can be grouped as: i) knowledge intensive; and ii) social regulation. These approaches have their advantages as well as disadvantages in terms of achieving their objectives and the sustainability of the initiatives (Table 7.1).

Knowledge-based Approach

The APFAMGS initiative is based on the principle of demystifying science through enhancing the capacities of the communities in terms of their skills and scientific knowledge. The focus is on facilitating or making communities assess the groundwater potential at the village level and estimating the available water before each crop season. These estimates are integrated at the hydrological unit level, providing the much needed scientific scale for assessing the groundwater. At the same time, the scale at which observation wells are monitored (village level) is more appropriate to the communities. For, official groundwater assessment is made based on the observation wells located at the Mandal (more than 30 villages) level and does not reflect the situation at the village level. Crop water budgets are prepared by the communities at the village level and the suggested cropping pattern for the season is provided (based on the groundwater availability) to the community. These details are shared across the villages within the hydrological unit.

The 'do it yourself' approach with relatively better scientific or technical inputs has clearly improved the awareness of the well owners. The initiative is highly successful in demystifying science and needs to be considered at the policy level to promote institutional linkages for generating such information at the village level. While such an awareness has helped in checking further expansion of groundwater development, i.e. new wells, it has failed to encourage other conservation practices such as increased investments in recharge structures or equity by sharing the water with un-irrigated farmers. Though our sample village does not provide any evidence on the reduction in water-intensive crops (paddy), it has been achieved in other places (Reddy, 2012). The limited impact is mainly due to the reason that neither social regulations are imposed, nor economic incentives are provided, for adopting such measures. In fact, the farmers feel that the APFAMGS merely plays an advisory role without any incentives or disincentives to follow the advisories. The result is that a lot of useful information is generated at the appropriate scale, helping only the well-owning farmers while the farmers hitherto not having wells are dissuaded from digging new wells (through information-based awareness) – there is no incentive for them to support the initiative; in fact, they are not even members of the committee.

Our qualitative research indicated that farmers are very much interested in having institutional arrangements in the lines of APFAMGS for managing groundwater. However, sustainability of the APFAMGS initiative is a big question mark in the absence of linkages with formal institutions, and policy or legislative backing of the movement*. Moreover, the exit protocol is not clearly defined. In a number of villages, the activities of the APFAMGS came to a standstill during the two years' gap (2009–11), due to the delay in the extension of the project.

Table 7.1

Features of the Three Institutional Models

Features	*APFAMGS*	*SRWM*	*APDAI*
Initiative (funding)	External (FAO)	External (AEI, Luxembourg)	State Government (DoRD)
Implementation	NGOs (BIRDS)	NGOs (CWS+Partners)	Govt.+NGO (WASSAN+Partners), Mahila Samkhyas
Years of existence**	8	7	2
Groundwater situation	Scarce	Scarce	Scarce
Project scale	Big (650 villages)	Small (19 villages)	Small (8 villages)
Key features	Information	Informal regulation	Formal regulation

Cont'd...

* Though HUNs are registered bodies and can take up activities like input procurement, output marketing, etc. they are yet to be functional in these activities.

** At the time of Field Survey.

...Cont'd

Scale of operation	Hydrological unit	Vicinity of wells (within a village)	Area based on the wells (within a village)
Institutional approach	Influencing community through generation of intensive scientific information	Regulating community through awareness and incentives	Regulating community through semi-scientific, information-based awareness and incentives
Operational modalities	All well owners with focus on information, followed an extensive approach	Small groups of well owners and dryland farmers, followed an intensive approach	Larger group of well owners and dryland farmers covering specific location, focus on incentives
Farmers' contribution	NIL	20 per cent towards micro-irrigation	75 per cent
Awareness on groundwater situation	High	High	High
Participation in management	Limited to well owners	High	High
Practising recommendations	Moderate	High	Low
Key to success	Professional approach	Leadership and incentives	Incentives
Impacts on access to water	Moderate	High	Moderate
Nature of key impact	Reduction in over-exploitation of groundwater	Conservation of water and sharing of water	Conservation and sharing of water
Impact on equity	No	Yes	Yes
Scalability	Good	Poor	Moderate
Sustainability	?	?	?

Source: Field Survey (PRA/FGD Methods) and Reports.

One suggestion made by the farmers in this regard is to bring the initiative under the groundwater department's purview so that the process would go on in the long run (Reddy et al., 2011).

Social Regulation Approach

The other two models, viz. the SRWM and APDAI, have adopted social regulation to manage groundwater. Though awareness building and data generation by the village communities are important components, the process is not so systematic. The most important aspect of these two models is to bring consensus among the communities to share water between well owners and others. Incentives such as reduced risk of well failure as no new wells are allowed, subsidies for micro-irrigation, provision for protective irrigation to the dry plots of the well owners, and the irrigation backup they get in the event of well failure, are put in place. Besides, there is provision for water harvesting structures to increase recharge, and distribution losses are reduced through pipeline supply of water and increased water use efficiency through promotion of micro-irrigation (subsidies).

Social regulation appears to be effective in terms of stopping new borewells as well as a larger number of households, especially the marginal and small, benefiting from sharing water with well owners. This not only helped in increasing the cropped area, but also provided protective irrigation to a number of plots during critical periods, thus saving the crops. This also resulted in equity in the distribution of water and overall improvement in welfare. However, there are differences between the two models of social regulation in terms of their effectiveness: the SRWM appears to be more effective when compared to APDAI. One reason could be that the SRWM was older, followed an intensive approach, and worked with smaller groups of farmers compared to the APDAI initiative. Though APDAI mostly follows the SRWM approach, it has adopted a broader (area based) and formal approach involving the department. Besides, groundwater management is one of the pilots under the APDAI and hence, there are chances of dilution as far as the departmental involvement is concerned.

Despite the formal approach, participation and rule following is limited in the APDAI. People indicated that there are no tangible benefits from the initiative, and 50 per cent of the farmers felt that the institutional arrangements are not feasible. This view is more conspicuous among those sharing wells. This sceptical nature could be due to the larger contribution (75%) from the farmers, which is substantial (total costs are ₹ 8 to 10 thousand per acres). On the other hand, the approach of peoples' contribution could provide the much needed ownership and sustainability.** It is observed that the formal process of entering an agreement with the witness of the *Tahsildar* has also discouraged some villages from joining the initiative.

The formal approach of APDAI appears good on paper, as it follows an integrated approach of drought adaptation. The integration also involves various departments such as rural development, groundwater, agriculture, etc. but the feasibility of such integration is doubtful. The approach involves the existing institutions such as the *Mahila Samakhyas*, which provide the assurance of sustenance in the medium run at least. However, at the same time, there is also a danger of acquiring the stamp of a Government programme where people look for freebies rather than regulation and contribution.

On the whole, the social regulation approach seems to work better for sustainable groundwater management when compared to the knowledge intensive approach. Water use and sharing through regulation has increased the area under protective irrigation in an equitable manner.

The knowledge intensive approach is not designed to address equity. In the absence of any regulations, formal or

---

** Of late people's contribution in Government programmes has lost its importance, as people are increasingly considering Government programmes as welfare measures rather than developmental. Hence, their contribution is treated as negative rather than as ownership.

informal, the farmers do not have any incentive to follow the good practices in the given policy environment. Encouraging sharing of water between well owners and others would result in achieving the twin objectives of conservation and improved access with equity. How to attain this on scale needs serious consideration at the policy level.

Sustainability of these initiatives is a major concern in all the approaches. None of the approaches have a well-defined exit protocol, while the APDAI appears to be well placed in this regard as its process involves a number of departments and formal institutions. At the same time, it requires strong leadership at the village level to implement and take the initiative forward, especially in the context of peoples' contribution. In the case of SRWM, its present success is mainly due to the commitment of NGO partners in the absence of any contribution from the farmers. Besides, in the absence of contribution, the financial sustainability of the initiatives would be a big concern, especially once the external funding stops. The weak sustainability of APFAMGS initiative was already evident during the no fund phase. Hence, fund flows appear to be critical for the success of the initiatives. The initiatives may continue in some of the villages due to strong leadership and commitment of the local NGOs even beyond the present funding, as they are at a smaller scale. Thus, scaling up these initiatives requires much more planning and designing.

Limitations of these Models

All these models suffer from limited scientific knowledge application at the ground level. The APFAMGS, which focuses on 'demystifying' science, does not follow a rigorous scientific approach towards groundwater recharge and balance estimation, water budgeting based on crop water requirement, etc. Similarly, the well-sharing and social regulation models do not integrate technical inputs for estimating the groundwater availability. Moreover, they do not consider scale impacts at a watershed or basin scale, as the positive impacts observed in the study locations may be causing negative impacts downstream.

Unless the impacts are considered at a scale of a hydrological unit, it is difficult to assess the real impacts.

Due to the short duration of these interventions, we are not able to provide hard core evidence to support some of the impacts that are measured in terms of farmers' perceptions. In the absence of long term data, the issue of attribution is also a problem. The changes in groundwater balance could be due to rainfall and other climatic fluctuations. Therefore, it is necessary to keep these limitations in view while considering scaling up of these initiatives.

Policy Directions for Scaling Up

Scientific information on geo-hydrology and groundwater is the domain of scientific community. Hydrology is treated as a pure physical science and hydrological information is often generated and disseminated in an esoteric form with little or no effort to bring it closer to the user communities. The technicalities involved in generating such information are believed to be beyond the knowledge of a non-technical person, not to mention the illiterate farmer.

Unlike other physical sciences, hydrology or hydrological information plays vital role in the day to day livelihoods of groundwater-dependent communities. The existing link between the scientific information and the users is very weak, serving no real purpose of helping the farming communities. Often, the information provided at a macro scale is inadequate and inappropriate to suit the micro-level situation and needs of the farmers. Thus, the scientific community should gear up to meet the needs of groundwater users through provision of more scientific and appropriate information to the users. The estimation methodologies need improvement along with increasing the number of observation wells and rain gauge stations. Policy makers should focus on providing hydrological information at a much lower scale than it is being done presently. Appropriate scale and methods suitable for hard rock areas as well as alluvial soils need to be developed. This becomes critical in the context of climate change. Policies should move towards focusing on groundwater management rather than development.

The increasing gap between the scientific information and the user, coupled with the fast deteriorating groundwater situation has led to institutional innovations of groundwater management. Some of the earlier institutional innovations have focused mainly on the community-based collective strategies such as forming rules and regulations for groundwater use and management (Deshpande and Reddy, 1990; World Bank, nd). None of the institutional arrangements based their approach on scientific information. Though some of them have achieved a fair amount of success, their spread and sustainability in the long run was limited, as they were driven by leadership and local conditions. Similarly, the regulator approaches of restricted power supply and no access to formal credit (to those who intend to have a borewell within a radius of 200 metres of another borewell) fail to encourage the farmers towards judicious use of groundwater.

For this purpose, innovative policies are needed, involving local communities and NGOs as partners. Generation of hydrological information at the village level is quite possible through the involvement of local communities and the NGOs. The NGOs can help in the process of capacitating the communities to take up the scientific activities. An assessment of three such models has provided useful insights for future policy.

The assessment of the three models indicates that CBGM is neither simple nor easily forthcoming. It calls for a lot of effort, working through complex rural dynamics at various levels. The reason is that appropriate policies to support or encourage such initiatives are not in place. Often, the existing policies work towards achieving opposite objectives rather than going in tandem with the participatory initiatives. The three approaches have proved that communities are capable of managing groundwater in a sustainable manner. The communities are also capable of understanding and using the technical aspects of hydro-geology. However, since groundwater is widely considered as a private property, there are no incentives for managing it at the community level. Furthermore, there are no economic incentives or disincentives for managing groundwater in a sustainable manner. Hence, unless wide-ranging policy changes are brought in, these initiatives remain as models rather than being adapted at a wider scale. Creating demand for these initiatives is as important as demand management of groundwater, and the demand management models cannot be effective as long as policy environment is supply-sided.

Some of the important policy interventions for promoting CBGM on a wider scale include:

- Need for dispelling the notion of groundwater as private property and making it a common property in the real sense. This calls for wide-ranging legislations and legal support.
- Establishing or moving towards community-based property rights on groundwater with proper distribution of rights to potential users.
- Moving towards aquifer planning at the hydrological unit level to start with and then to watershed or river basin scale.
- Creating hydrological information at a much smaller scale appropriate for short-term farming decisions. This could be attained through creating low cost infrastructure at the village level and providing training at the local level to take up the responsibilities on a regular basis with the necessary economic incentives.
- Water sharing at the village level needs to be promoted as a first step in this direction. Existing wells could be linked and termed as common property.
- Incentives to conserve and manage water resources rather than exploiting the resources such as free power and support prices for water-intensive crops like paddy.
- The present policy distortions of free power, and the input and output pricing policies need to be rationalised to match conservation objectives.
- Regulation through pricing is the most effective instrument, but is hardly adopted at the policy level. In the

absence of realistic pricing, water use efficiency remains a dream.

- As long as water rights are linked to land, water sharing is the best option to achieve equity. Encouraging and strengthening the existing traditional group wells in AP through differential and higher incentives in electricity tariffs, subsidies for micro-irrigation kits, etc. would help improve the equity and sustainability of groundwater.
- Andhra Pradesh Water Land and Trees Act (APWALTA) bans drilling of new wells in villages notified as over-exploited. The Government may encourage only new wells on group sharing basis in villages/micro-basins that are identified as critical and semi-critical with respect to groundwater development. Strengthening and enforcing the existing regulations like APWALTA could be a starting point in this direction.
- Delinking land and water rights needs to be treated as an important policy goal, at least in the long run.

Thus, the experience of the three models reveals that wide ranging policy changes are required to scale up the achievements of these small scale initiatives. Replication of these models could be possible with high transaction costs, but the sustainability of these initiatives remains uncertain in the present policy environment.

The most important lessons from these models include:

1. creation of information at the appropriate scale through community involvement; and
2. generating demand for demand management of groundwater with the help of this information.

However, the conclusions drawn here are based on the experience of a few villages and hence cannot be generalised. While these findings provide some insights, there is a need for better understanding of such initiatives through a large scale systematic research covering the existing initiatives across the country.

Annexures

Table 1

Districts by Category of Groundwater Development in AP

Category	*No. of Districts Falling Under Different Categories*															
	1985			*1993*	*2002*			*2004*			*2007*			*2009*		
	C	*NC*	*Overall*	*Overall*	*C*	*NC*	*Overall*	*C*	*NC*	*Overall*	*C*	*NC*	*Overall*	*C*	*NC*	*Overall*
Safe	22	18	22	22	18	15	16	17	9	17	19	12	18	18	15	16
Semi-Critical	0	1	0	0	0	4	5	2	10	5	0	8	2	1	7	6
Critical	0	2	0	0	0	2	1	0	3	0	0	1	2	0	0	0
Over-Exploited	0	1	0	0	0	1	0	0	0	0	0	1	0	0	0	0
NA	1	1	1	1	5	1	1	4	1	1	4	1	1	4	1	1
Total	23	23	23	23	23	23	23	23	23	23	23	23	23	23	23	23

Source: Data compiled from Groundwater Resource Estimated Reports different years, Ground Water Department, GoAP.

Note: 1. C=Command (*ayacut*); NC=Non-Command (non-*ayacut*); Total=Overall (both command and non-command areas together). NA= Not applicable or available.

2. Area under command in each district is considered and hence total number of districts turns out to be more.

Table 2

District-wise Distribution of Over-Exploited (OE) Villages in OE Basins and Number of Failed Wells

District/ Region	2002		2005	2008		2009	% of Assessment Units Reporting Repeated Depletion up to 2008	% of Assessment Units Reporting Repeated Depletion up to 2009
	No. of OE Villages	No. of Failed Wells	No. of OE Villages	No. of OE Villages	No. of Failed Wells	No. of OE Villages		
Srikakulam	0	2,977 (384)	0	0	2,598 (188)	11	2.6	0
Vizianagaram	0	6,872 (739)	4	0	6,161 (302)	0	0	0
Visakhapatna m	4	4,106 (382)	0	0	2,961 (259)	0	4.7	0
East Godavari	0	705 (488)	32	0	921 (642)	0	5.3	0
West Godavari	132	1,251 (563)	152	118	1,001 (809)	0	23.9	0
Krishna	0	4,733 (1906)	34	40	2,969 (571)	36	10	4

Cont'd...

...Cont'd

Guntur	0	1,720 (199)	11	4	3,899 (513)	3	0	0
Prakasam	28	9,386 (601)	145	121	11,907 (5047)	98	26.8	16.1
Nellore	19	7,532 (1,117)	150	123	10,449 (1,070)	0	8.7	0
Chittoor	89	24,396 (1,382)	569	601	41,986 (2,880)	422	51.5	33.3
Cuddapah	499	23,141 (919)	345	379	2,8191 (1,525)	202	68.6	11.8
Anantapur	133	19,031 (962)	385	420	24,476 (3,128)	271	68.3	42.9
Kurnool	5	6,222 (538)	106	46	5,667 (1,501)	16	9.3	0
Mahbubnagar	36	51,723 (2,554)	378	158	45,495 (9,360)	73	48.4	7.8
Ranga Reddy	112	12,328 (2,458)	390	332	10,431 (1,737)	108	37.8	8.1

Cont'd...

...Cont'd

Medak	45	25,491 (1,976)	380	247	9,840 (2,743)	293	47.8	26.1
Nizamabad	150	4,099 (923)	196	231	1,534 (482)	167	75	25
Adilabad	0	3,321 (416)	47	77	1,909 (344)	6	3.8	0
Karimnagar	38	12,492 (491)	294	153	115 (16)	4	45.6	0
Warangal	125	12,230 (825)	330	208	13,339 (4,446)	312	44	44
Khammam	9	5,290 (349)	34	16	1,918 (248)	23	8.7	2.2
Nalgonda	57	21,017 (1,065)	208	175	6,220 (1,786)	78	50.8	11.9
Andhra Pradesh	1,481	26,0063 (21,237)	4,190	3,449	23,3987 (3,9597)	2123	30.3	11.3

Note: Figures in the brackets indicate the total no. of failed tubewells

Source: 1. OE Villages Data (data compiled from): GoAP, Groundwater Resource Estimated years Reports, Ground Water Department, Hyderabad

2. Failed Well Data (data compiled from): GoAP, 3rd and 4th Minor Irrigation Census, Directorate of Economics and Statistics, Hyderabad

Table 3

Descriptive Statistics of Selected Variables (N=22)

Variables	*1985*		*1993*		*2002*		*2004*		*2007*		*2009*	
	Mean	*SD*	*Mean*	*SD*	*Mean*	*SD*	*Mean*	*SD*	*Mean*	*SD*	*Mean*	*SD*
AARF (in mm)	731.14	194.02	814.36	150.17	919.05	158.79	744.95	180.67	861.12	270.23	855.41	203.60
I I	126.61	19.18	127.62	18.99	128.97	19.81	127.91	19.50	133.22	20.52	136.89	20.34
CI	118.02	14.73	125.31	19.68	124.77	19.89	124.37	19.28	129.01	21.95	130.35	20.99
HDI	–	–	0.39	0.07	0.62	0.10	0.52	0.07	0.52	0.07	0.52	0.07
% Persons BPL/HPI*	55.09	15.10	40.82	11.93	34.19	12.93	31.33	13.72	28.60	14.73	0.49	0.05
PCI (in ₹)	–	–	7,319.00	1,264.49	9,934.50	1,697.68	19,167	4,314.19	22,845.27	4,787.88	22,845.27	4,787.88
Literacy Levels	27 69	6.37	41.64	7.25	58.69	7.00	58.69	7	58.69	7	58.69	7.00
Groundnut (in ha)	15,019.27	12,716.8	19,475.7	15,804.1	13,631.0	13,592.5	11,634.9	12,046.3	11,276.77	12,230.58	13,359.18	16,392.37

Cont'd...

...Cont'd

No. of Dugwells	2,117.91	1,933.52	937.36	896.06	513.05	696.95	546.77	1,130.51	161.09	353.18	161.09	353.18
No. of Tubewells	614.09	556.87	1,302.77	1,210.13	1,291.23	1,147.41	2,114.59	2,354.85	594.50	651.87	594.50	651.87
No. of Agrl. Service Connections	2,011	1,869.67	4,644.36	2,718.03	5,374	3,475.52	2,943.64	2,110.43	3,953.50	2,239.61	121803.09	77810.61
No. of WSD	–	–	–	–	303.50	229.04	435.36	358.39	503.77	403.62	358.77	342.66
No. of Tanks	3,590.50	2,770.95	3,632.50	2,810.64	3,747.41	2,786.43	3,747.41	2,786.43	2,717.68	2,341.15	2,717.68	2,341.15
NIA Canals (in ha)	81,540.41	92,234.2	75,409.8	84,299.5	74,972.1	81,089.5	61,180.8	76,910.4	73,761.55	80,993.86	75,883.95	81,796.50

* HPI is used in 2009.

Table 4

Factors Influencing Groundwater Development in Command and Non-Command Areas (1985)

Variable	*Command*		*Non-Command*		*Overall*	
	Coefficient	*VIF*	*Coefficient*	*VIF*	*Coefficient*	*VIF*
Constant	-33.18 (-2.96)*		-50.741 (-2.033)**		-55.357 (-2.587)*	
I I	0.096 (1.84)**	1.105	–	–	–	–
No. of Tubewells	0.002 (1.27)	1.120	–	–	–	–
% of BPL Population	0.161 (2.14)*	1.548	–	–	0.399 (2.257)*	1.572
Literacy Levels	0.564 (2.868)*	1.986	2.177 (3.021)*	1.606	1.379 (3.000)*	1.995
No. of Agrl. Service Connections	0.002 (2.412)*	1.491	0.015 (5.361)*	1.683	0.009 (5.664)*	1.490
No. of Tanks	–	–	0.004 (2.184)*	1.373	0.001 (1.094)	1.456
Groundnut	–	–	-0.001 (-1.762)**	1.233	–	-
R^2 (R Bar)	0.57 (0.41) N=22		0.73 (0.66) N=22		0.74 (0.67) N=22	

Note: Figures in the brackets refer to the't' values and *, ** indicate level of significance at 5 and 10 per cent respectively.

Table 5

Factors Influencing Groundwater Development in Command and Non-Command Areas (1993)

Variable	*Command*		*Non-Command*		*Overall*	
	Coefficient	*VIF*	*Coefficient*	*VIF*	*Coefficient*	*VIF*
Constant	-15.76 (-1.14)	–	-15.77 (-1.14)	–	-15.63 (-1.132)	–
II	0.186(1.790)**	1.559	0.185(1.776)**	1.559	0.184 (1.775)**	1.559
No. of Dugwells	0.003 (1.345)	1.115	0.003(1.342)	1.115	0.003(1.346)	1.115
No. of Tubewells	0.002(1.217)	1.521	0.002(1.224)	1.521	0.002(1.218)	1.521
HDI	47.342(1.919)**	1.348	47.738(1.931)**	1.348	47.434(1.920)**	1.348
NIA Canals	-0.000 (-3.930)*	1.492	-0.000 (-3.926)*	1.492	-0.000 (-3.931)*	1.492
R^2 (R Bar)	0.65 (0.55) N=22		0.65 (0.55) N=22		0.65 (0.55) N=22	

Note: Figures in the brackets refer to the't' values and *, ** indicate level of significance at 5 and 10 per cent respectively.

Table 6

Factors Influencing Groundwater Development in Command and Non-Command Areas (2002)

Variable	*Command*		*Non-Command*		*Overall*	
	Coefficient	*VIF*	*Coefficient*	*VIF*	*Coefficient*	*VIF*
Constant	-4.227 (-0.178)		-72.546 (-1.693)	–	-41.885 (-1.097)	–
AARF	–	–	–	–	-0.040 (-1.775)**	1.491
II	-0.061 (-0.474)	1.544	0.617 (2.512)*	1.416	–	–
No. of Dugwells	-0.006 (-1.859)**	1.361	-0.003 (-0.527)	1.177	–	–
No. of Tubewells	0.008 (2.466)*	1.916	0.011 (2.572)*	1.547	–	–
Crop Intensity	–	–	–	–	.479 (2.390)*	1.841
HDI	24.605 (0.795)	1.875	51.600 (1.144)	1.232	110.886 (3.377)*	1.265
% of BPL Pop.	0.675 (3.320)*	1.523	0.178 (0.489)	1.325	–	–
No. of Watersheds	-0.038 (-2.890)*	2.026	-0.006 (-0.252)	2.026	–	–
NIA Canals	–	–	–	–	0.000 (-4.792)*	1.818
Groundnut	–	–	–	–	0.001 (2.409)*	1.870
R^2 (R Bar)	0.60 (0.40) N=22		0.64 (0.49) N=22		0.76 (0.68) N=22	

Note: Figures in the brackets refer to the't' values and *, ** indicate level of significance at 5 and 10 per cent respectively.

Table 7

Factors Influencing Groundwater Development in Command and Non-Command Areas (2004)

Variable	*Command*		*Non-Command*		*Overall*	
	Coefficient	*VIF*	*Coefficient*	*VIF*	*Coefficient*	*VIF*
Constant	36.760 (2.091)**	–	177.744 (6.261)*	–	35.896 (1.435)	–
AARF	–	–	-0.115 (-4.502)*	1.727	-0.040 (-2.275)*	1.219
II	–	–	–	–	0.482 (3.236)*	1.023
No. of Tubewells	0.007 (2.245)*	1.930	0.004 (2.581)*	1.356	0.003 (2.366)*	1.143
PCI (constant prices)	0.001 (1.190)	1.286	0.000 (-0.322)	1.238	–	–
No. of Watersheds	-0.034 (-3.090)*	2.114	-0.038 (-2.667)*	2.132	–	–
No. of Tanks	-0.001 (-0.576)	1.379	-0.003 (-2.262)*	1.362	-0.003 (-2.457)*	1.291
NIA Canals	0.000 (-5.342)*	1.512	0.000 (-0.881)	1.444	0.000 (-4.895)*	1.361
R^2 (R Bar)	0.71 (0.60) N=22		0.71 (0.60) N=22		0.81 (0.75) N=22	

Note: Figures in the brackets refer to the't' values and *, ** indicate level of significance at 5 and 10 per cent respectively.

Table 8

Factors Influencing Groundwater Development in Command and Non-Command Areas (2007)

Variable	*Command*		*Non-Command*		*Overall*	
	Coefficient	*VIF*	*Coefficient*	*VIF*	*Coefficient*	*VIF*
Constant	161.492 (3.341)*		-38.909 (-1.047)	–	67.707 (2.600)*	–
AARF	–	–	-.074 (-4.982)	1.603	-0.062 (-4.226)*	1.571
II	–	–	–	–	0.394 (2.425)*	1.115
No. of Dugwells	0.007 (.609)	1.19	–	–	–	–
No. of Tubewells	–	–	–	–	-0.011 (-1.774)**	1.515
Crop Intensity	–	–	0.735 (3.500)*	2.112	–	–
HDI	-192.377 (-2.5)*	2.05	–	–	–	–
PCI	–	–	–	–	0.000 (-0.076)	1.104
Literacy Levels	–	–	1.173 (2.207)*	1.376	–	–
No. of Agrl. Service Connections	–	–	0.003 (1.799)**	1.151	–	–
No. of Watersheds	-0.029 (-2.316)*	2.12	–	–	–	–
No. of Tanks	-0.006 (-2.823)*	1.76	–	–	–	–
NIA Canals	0.000 (-1.909)**	1.66	0.000 (-3.198)*	1.790	0.000 (-4.556)*	1.095
R^2 (R Bar)	0.49 (0.34) N=22		0.74 (0.67) N=22		0.74 (0.67) N= 22	

Note: Figures in the brackets refer to the't' values and *, ** indicate level of significance at 5 and 10 per cent respectively.

Table 9

Factors Influencing Groundwater Development in Command and Non-Command Areas (2009)

Variable	Command		Non-Command		Overall	
	Coefficient	*VIF*	*Coefficient*	*VIF*	*Coefficient*	*VIF*
Constant	141.876 (4.342)*		103.749 (3.050)*		126.017 (3.442)*	–
AARF	–	–	–	–	–0.014 (–0.925)	1.281
II	–	–	–	–	–	–
No. of Dugwells	0.019 (2.674)*	1.127	–	–	–	–
No. of Tubewells	–	–	0.008 (1.309)	1.207	–	–
Crop Intensity	–	–	–	–	–	–
HDI	–	–	–	–	–	–
PCI	–	–	–	–	–	–
HPI	–180.717 (–2.887)*	1.588	–151.649 (–2.188)*	1.129	–147.816 (–2.573)*	1.272
No. of Agrl. Service Connections	–	–	0.000 (3.282)*	1.095	0.000 (2.845)*	1.376
No. of Watersheds	–0.015 (–1.827)**	1.311	–	–	–	–
No. of Tanks	–	–	–	–	–	–
NIA Canals	0.000 (–4.915)*	1.680	–	–	0.000 (–3.076)*	1.559
R2 (R Bar)	0.76 (0.70) N= 22		0.53 (0.46) N= 22		0.72 (0.65) N= 22	

Note: Figures in the brackets refer to the 't' Values and *, ** indicate level of significance at 5 and 10 per cent respectively.

Table 10

Factors Influencing Groundwater Development in DPAP and Non-DPAP Regions (1985)

Variable	*DPAP*			*Non-DPAP*		
	Coefficient	*t*	*VIF*	*Coefficient*	*t*	*VIF*
Constant	61.999	3.499		-30.675	-1.551	
No. of Dugwells	–	–	–	0.006*	7.498	1.069
II	–	–	–	0.225	1.961	1.085
AARF	-0.049**	-1.937	1.062	0.025**	2.230	1.112
No. of Agrl. Service Connections	0.004**	2.083	1.049	–	–	–
NIA under Project Canals	0.000*	-2.496	1.034	–	–	–
R^2 (R Bar)	0.65 (0.5) N= 12			0.92 (0.87) N= 10		

Note: *, ** indicate level of significance at 5 and 10 per cent respectively.

Table 11

Factors Influencing Groundwater Development in DPAP and Non-DPAP Regions (1993)

Variable	*DPAP*			*Non-DPAP*		
	Coefficient	*T*	*VIF*	*Coefficient*	*t*	*VIF*
Constant	-15.249	-0.732		-0.495	-0.056	
CI	0.462*	2.671	1.388	–	–	–
AARF	-0.032*	-3.328	1.113	–	–	–
NIA Canals	0.000*	-4.838	1.394	–	–	–
Literacy Level	0.594*	2.502	1.131	–	–	–
No. of Agrl. Service Connections	–	–	–	0.003*	2.892	1.245
% of Population BPL	–	–	–	0.392	1.501	1.245
R^2	0.85 (0.77) N= 12			0.72 (0.64) N= 10		

Note: *, ** indicate level of significance at 5 and 10 per cent respectively.

Table 12

Factors Influencing Groundwater Development in DPAP and Non-DPAP Regions (2002)

Variable	*DPAP*			*Non-DPAP*		
	Coefficient	*t*	*VIF*	*Coefficient*	*t*	*VIF*
Constant	-89.807	-1.209		-151.371	-3.033	
NIA Canals	-0.001*	-5.294	1.375	–	–	–
% of Population BPL	-0.558	-1.433	1.627	1.317*	4.953	1.321
CI	1.446*	2.307	1.208	–	–	–
Coverage of WSDP	0.060**	2.236	1.439	–	–	–
AARF	–	–	–	0.115**	2.225	1.791
No. of Tubewells	–	–	–	0.010**	2.078	1.409
II	–	–	–	0.168	1.386	1.249
R^2	0.82 (0.71) N= 12			0.91 (0.84) N= 10		

Note: *, ** indicate level of significance at 5 and 10 per cent respectively.

Table 13

Factors Influencing Groundwater Development in DPAP and Non-DPAP Regions (2004)

Variable	*DPAP*			*Non-DPAP*		
	Coefficient	*t*	*VIF*	*Coefficient*	*t*	*VIF*
Constant	138.847	2.013		137.115	2.465	
AARF	-0.067*	-3.236	1.418	–	–	–
II	0.553	1.618	1.621	0.350**	2.158	1.032
HPI (2001)	-195.221*	-2.811	1.405	-198.943**	-2.140	1.497
NIA Canals	0.000**	-1.939	1.438	0.000*	-3.506	1.665
PCI	–	–	–	-0.001	-1.714	1.196
R^2	0.90 (0.84) N= 12			0.84 (0.72) N= 10		

Note: *, ** indicate level of significance at 5 and 10 per cent respectively.

Table 14

Factors Influencing Groundwater Development in DPAP and Non-DPAP Regions (2007)

Variable	*DPAP*			*Non-DPAP*		
	Coefficient	*t*	*VIF*	*Coefficient*	*t*	*VIF*
Constant	129.822	6.663		-3.262	-0.130	
AARF	-0.055*	-3.045	1.211	–	–	–
No. of Agrl. Service Connections	-0.005**	-1.964	1.212	0.004*	2.787	1.002
NIA Canals	0.000*	-3.241	1.166	0.000*	–3.274	1.001
Irrigation Intensity	–	–	–	0.263	1.586	1.001
R^2	0.76 (0.67) N= 12			0.79 (0.68) N= 10		

Note: *, ** indicate level of significance at 5 and 10 per cent respectively.

Table 15

Factors Influencing Groundwater Development in DPAP and Non-DPAP Regions (2009)

Variable	*DPAP*			*Non-DPAP*		
	Coefficient	*t*	*VIF*	*Coefficient*	*t*	*VIF*
Constant	–81.445	–1.264		80.527*	3.017	
No. of Agrl. Service Connections	–	–	–	0.000**	2.351	1.536
Coverage of WSDP	0.034**	2.274	1.381	–0.119**	–2.278	1.377
NIA Canals	0.000*	–3.904	1.102	0.000**	–2.041	1.431
PCI	–	–	–	–0.001	–1.572	1.474
CI	1.136**	2.188	1.499	–	–	–
R^2	0.71 (0.60) N= 12			0.86 (0.75) N= 10		

Note: *, ** indicate level of significance at 5 and 10 per cent respectively

Figure 1

Districts Falling Under Different Categories in AP

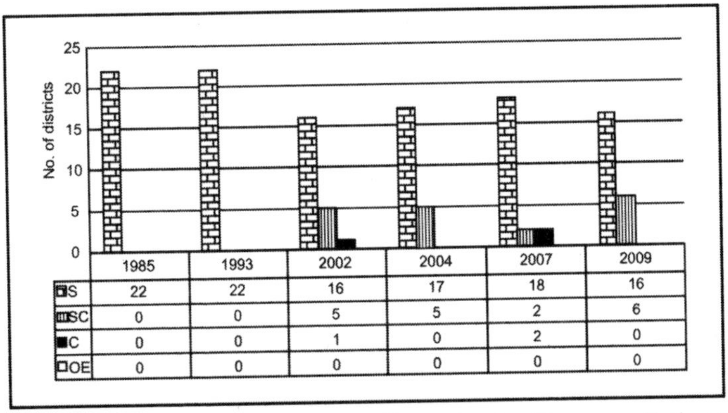

	1985	1993	2002	2004	2007	2009
S	22	22	16	17	18	16
SC	0	0	5	5	2	6
C	0	0	1	0	2	0
OE	0	0	0	0	0	0

Source: Data compiled from GoAP. Groundwater Resource Estimated Reports different years, Ground Water Department, Hyderabad

Note: S=Safe; SC=Semi-Critical; C=Critical; OE=Over-Exploited

Figure 2

Districts Under Different Categories Across Command and Non-Command Areas in AP

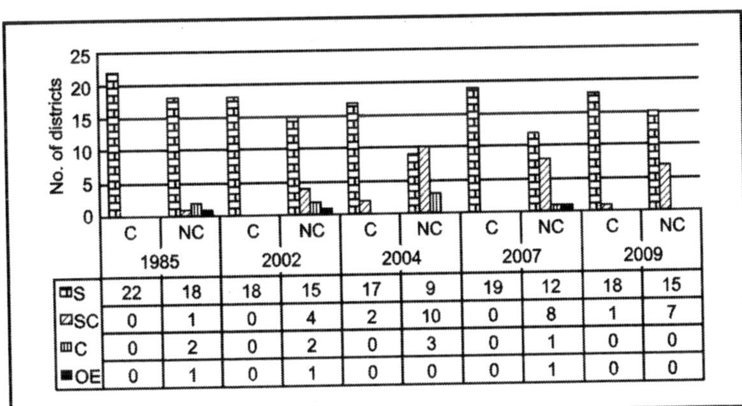

	1985		2002		2004		2007		2009	
	C	NC	C	NC	C	NC	C	NC	C	NC
S	22	18	18	15	17	9	19	12	18	15
SC	0	1	0	4	2	10	0	8	1	7
C	0	2	0	2	0	3	0	1	0	0
OE	0	1	0	1	0	0	0	1	0	0

Source: Data compiled from GoAP. Groundwater Resource Estimated Reports different years, Ground Water Department, Hyderabad

Note: S=Safe; SC=Semi-Critical; C=Critical; OE=Over-Exploited

Map 1

Rural Poverty, Drought-Prone Areas, Agro-climatic Zones. River Basins and Forest Types Integrated Map of India

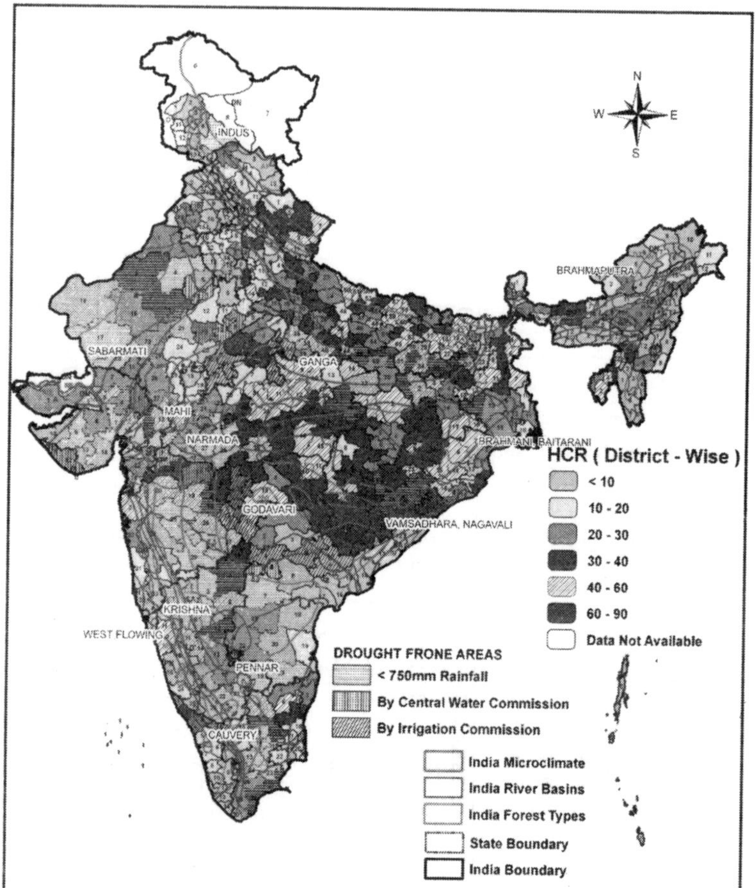

Map 2

Physiographical Map of Andhra Pradesh

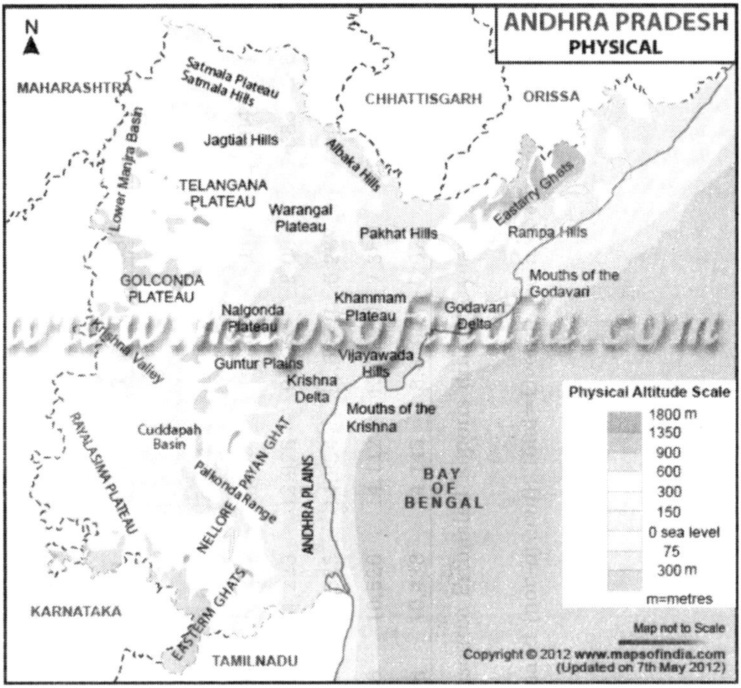

Map 3

Agro-Climatic Zones of Andhra Pradesh

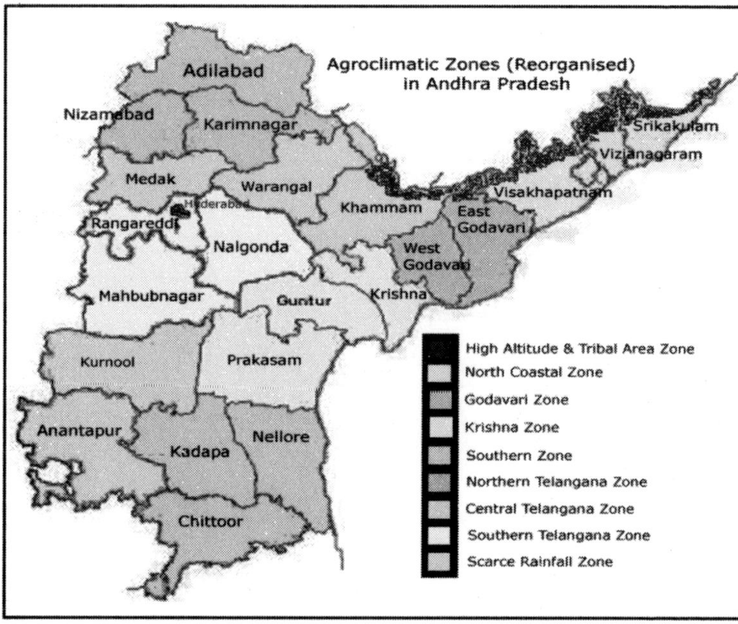

Source: Acharya N G Ranga Agricultural University (2008), Annual Report 2007–08, Hyderabad.

Map 4

Hydrogeology Map of Andhra Pradesh

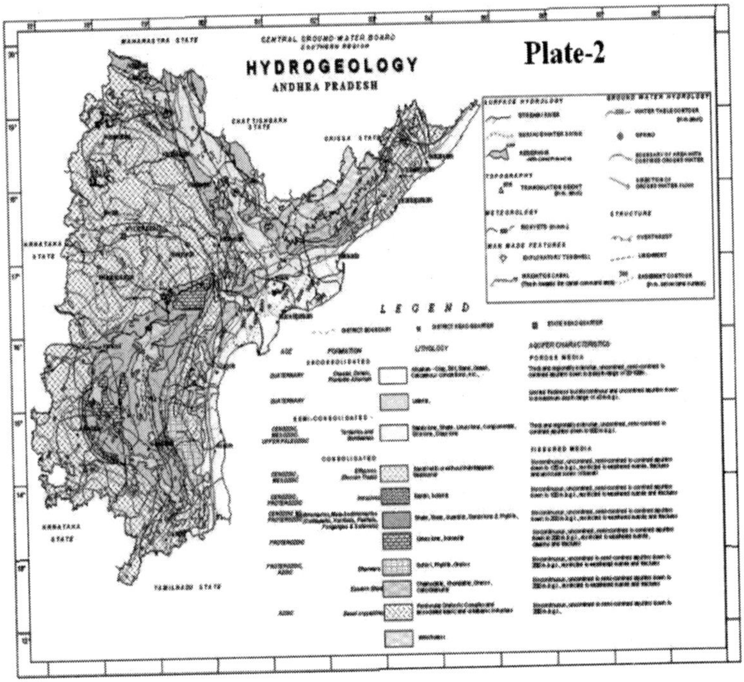

Map 5

Geological Map of Andhra Pradesh

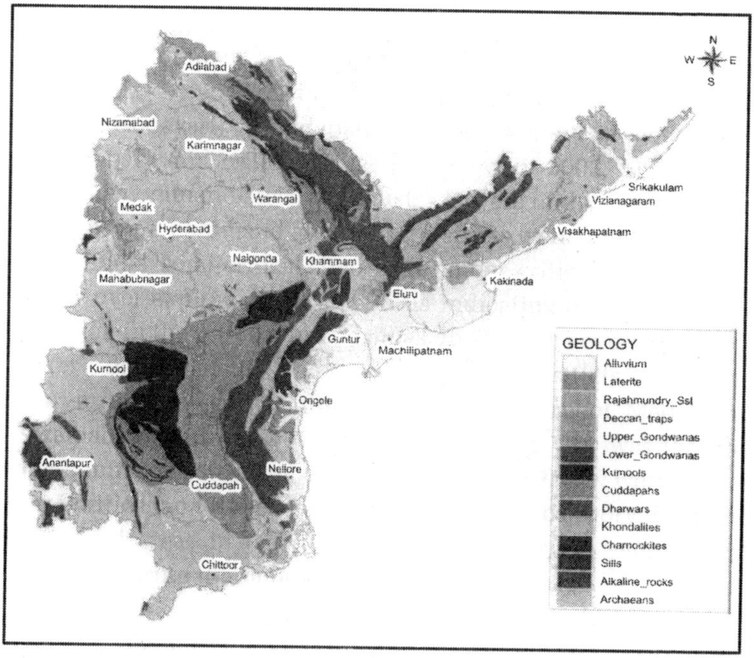

Source: GoAP (2008), Groundwater Resource Andhra Pradesh 2007, Vol-I, Department of Ground Water, August. Hyderabad.

Map 6

Soils of Andhra Pradesh

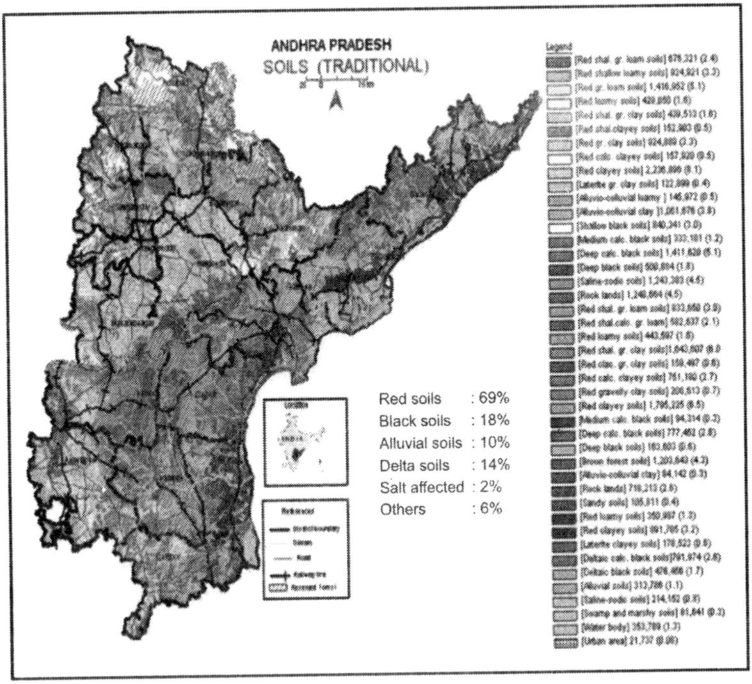

Map 7

Status of Groundwater Development in Andhra Pradesh – 1985 (Mandal wise)

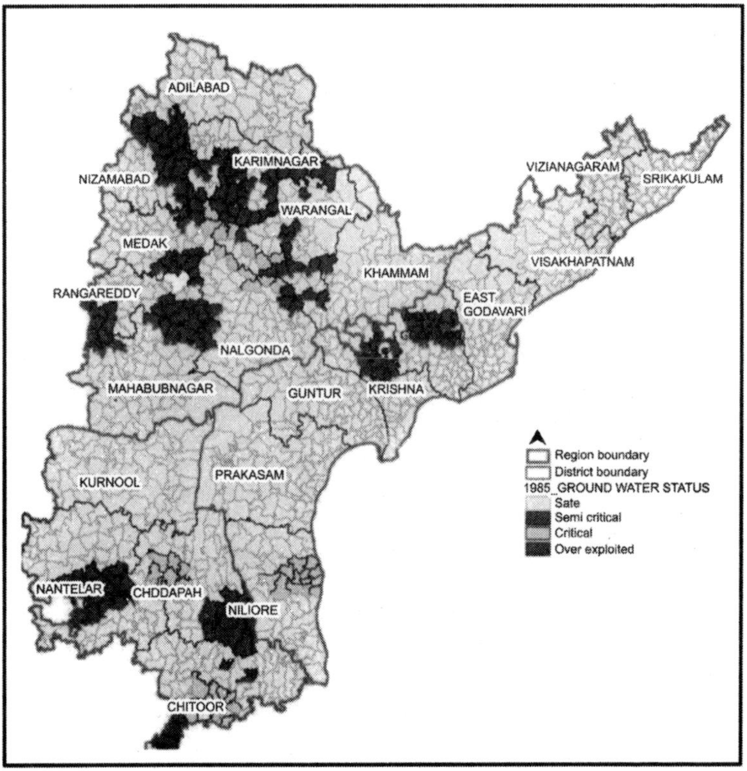

Map 8

Status of Groundwater Development in Andhra Pradesh – 1993 (Mandal wise)

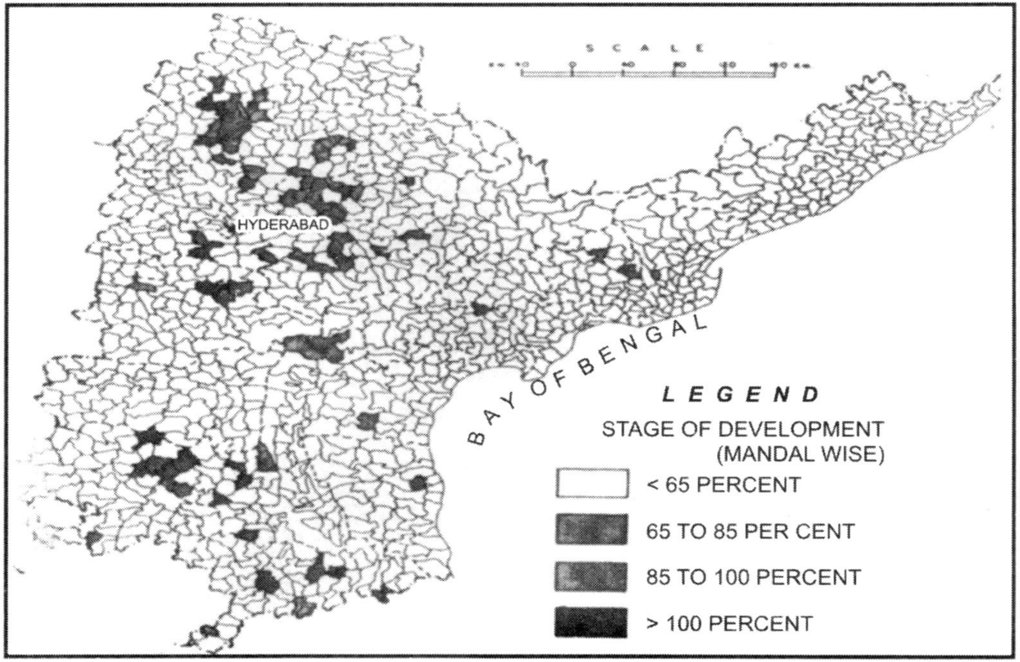

Map 9

Status of Groundwater Development in Andhra Pradesh – 2002 (Basin wise)

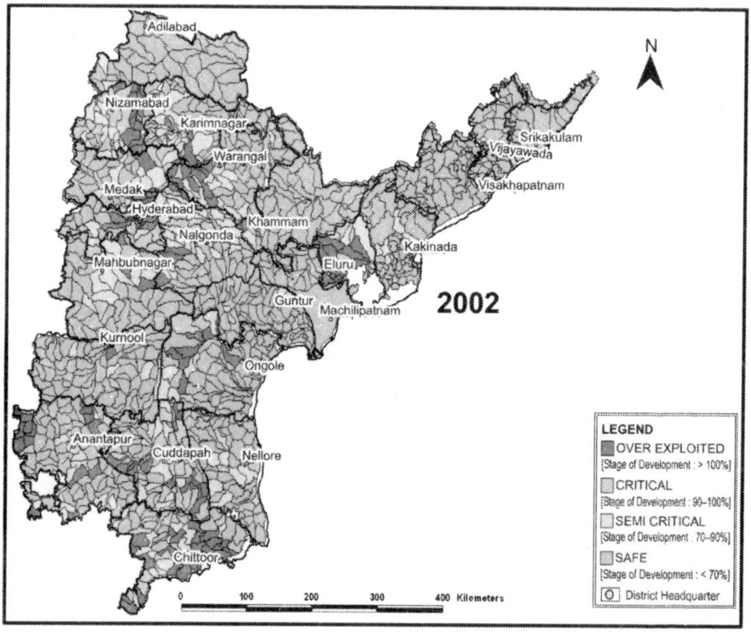

Map 10

Status of Groundwater Development in Andhra Pradesh – 2002 (Mandal wise)

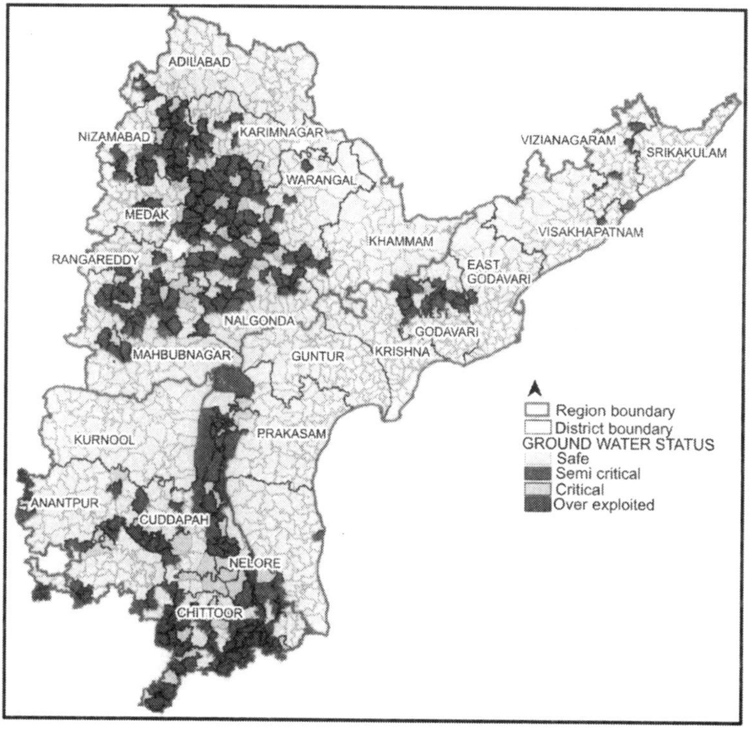

Map 11

Status of Groundwater Development in Andhra Pradesh – 2004 (Basin wise)

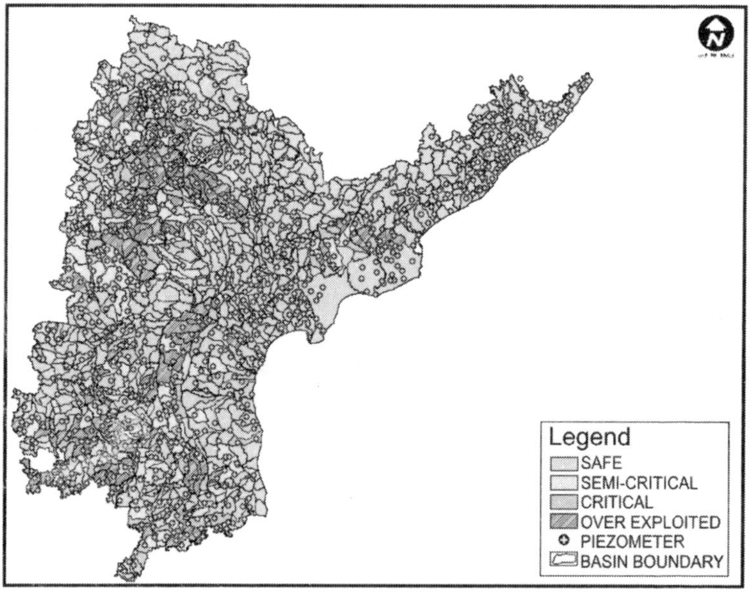

Map 12

Status of Groundwater Development in Andhra Pradesh – 2004 (Mandal wise)

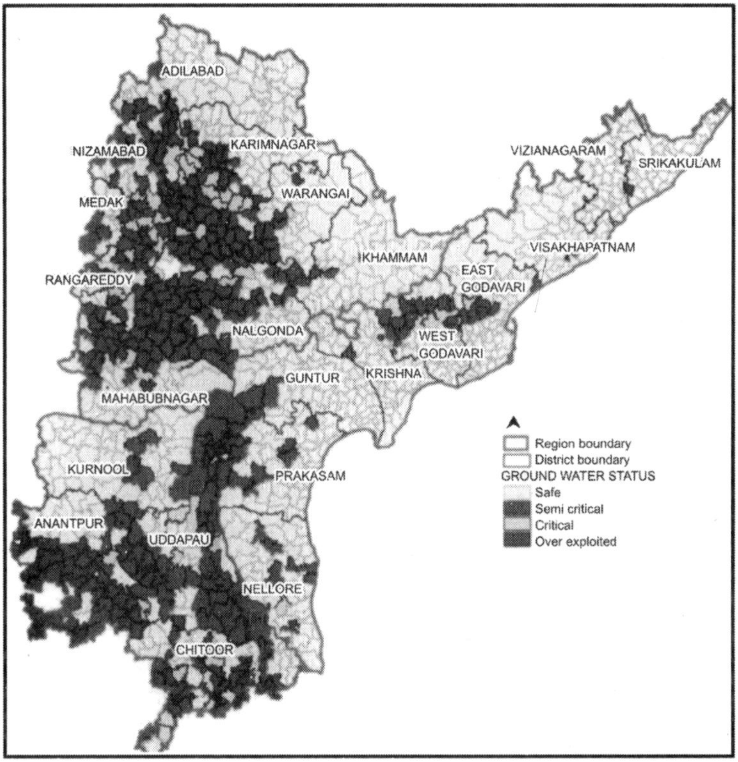

Map 13

Status of Groundwater Development in Andhra Pradesh – 2007 (Basin wise)

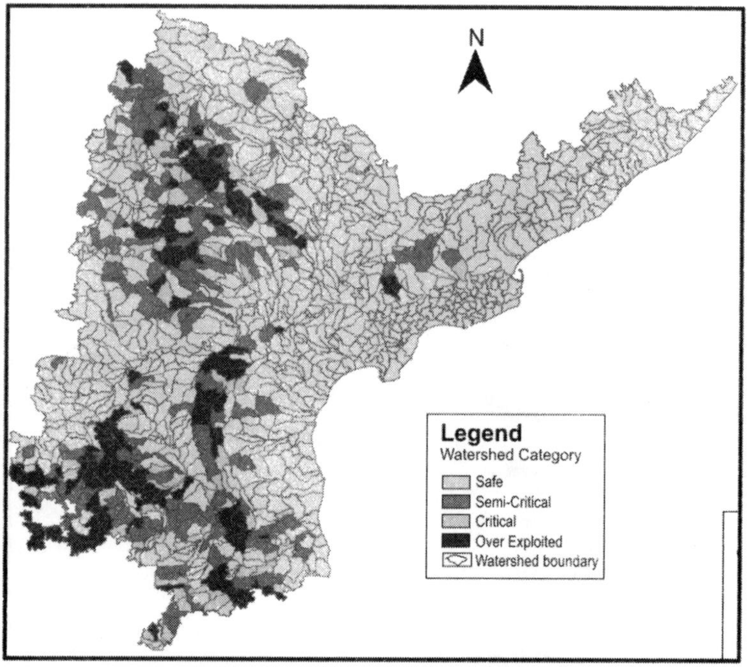

Source: GoAP (2008), Groundwater Resource Andhra Pradesh 2007, Vol I, Department of Ground Water, Hyderabad, August.

Map 14

Status of Groundwater Development in Andhra Pradesh – 2007 (Mandal wise)

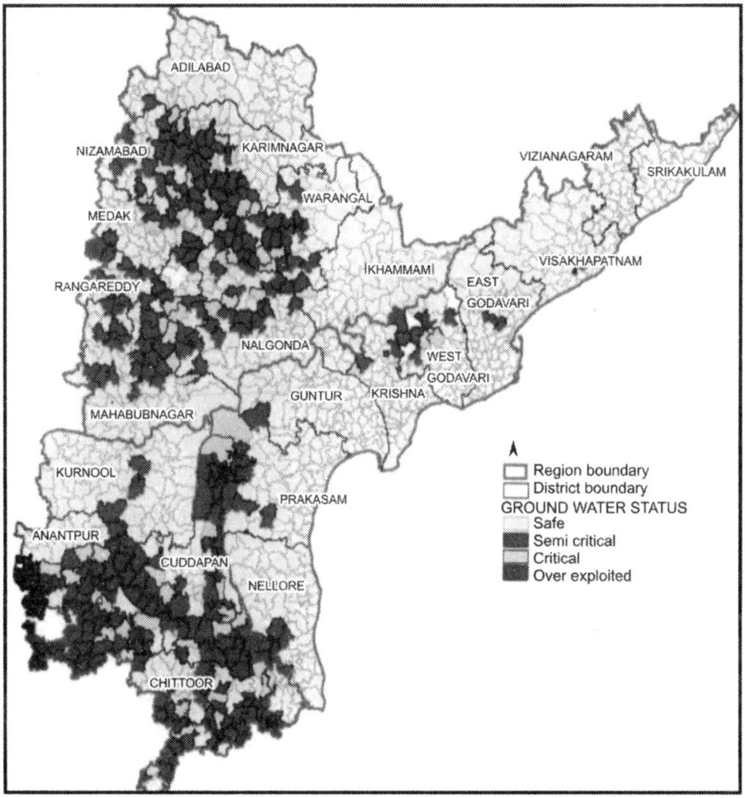

Map 15

Status of Groundwater Development in Andhra Pradesh – 2009 (Basin wise)

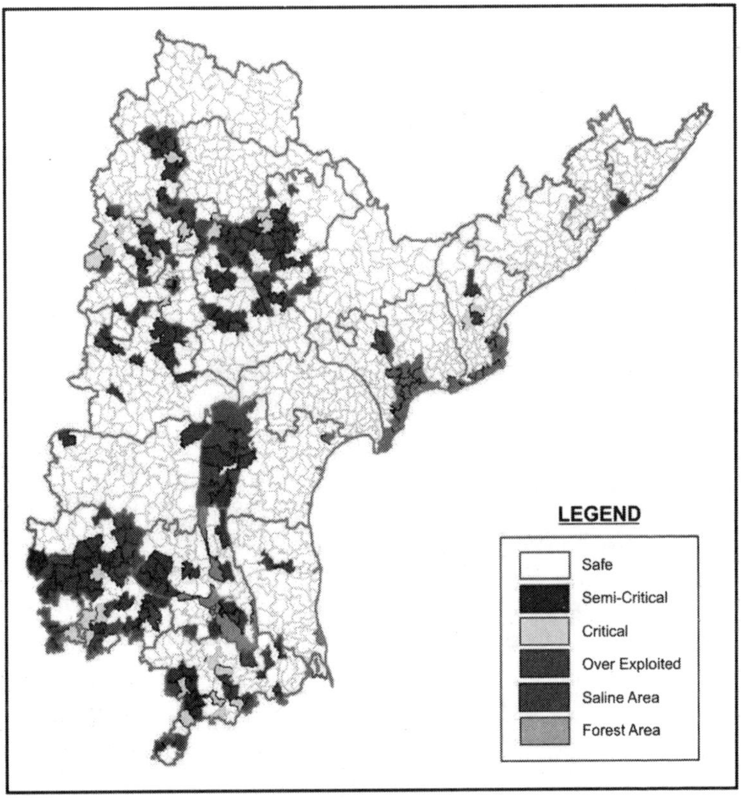

Map 16

Status of Groundwater Development in Andhra Pradesh – 2009 (Mandal wise)

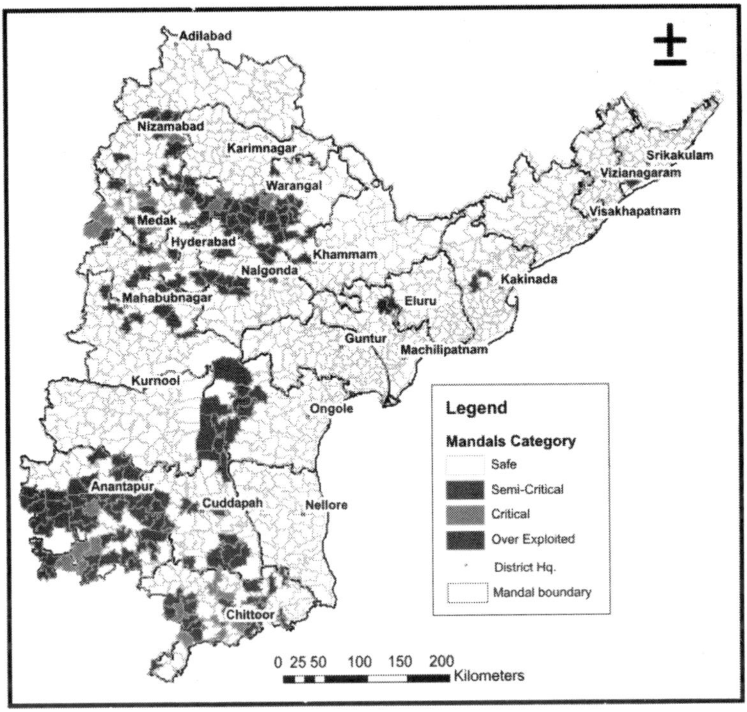

Map 17

Status of Groundwater Development in Andhra Pradesh – 2009
(Watershed wise)

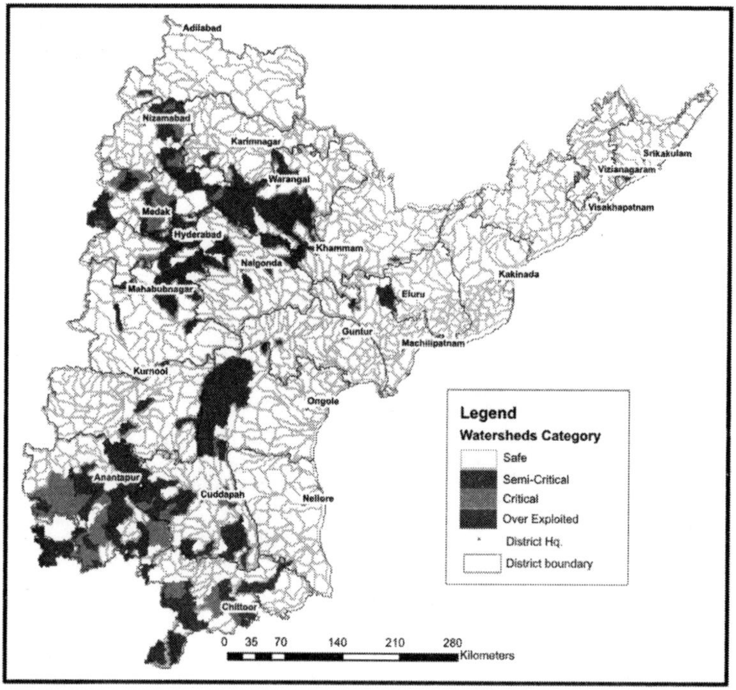

Map 18

DPAP/Non-DPAP Mandals/Districts in Andhra Pradesh

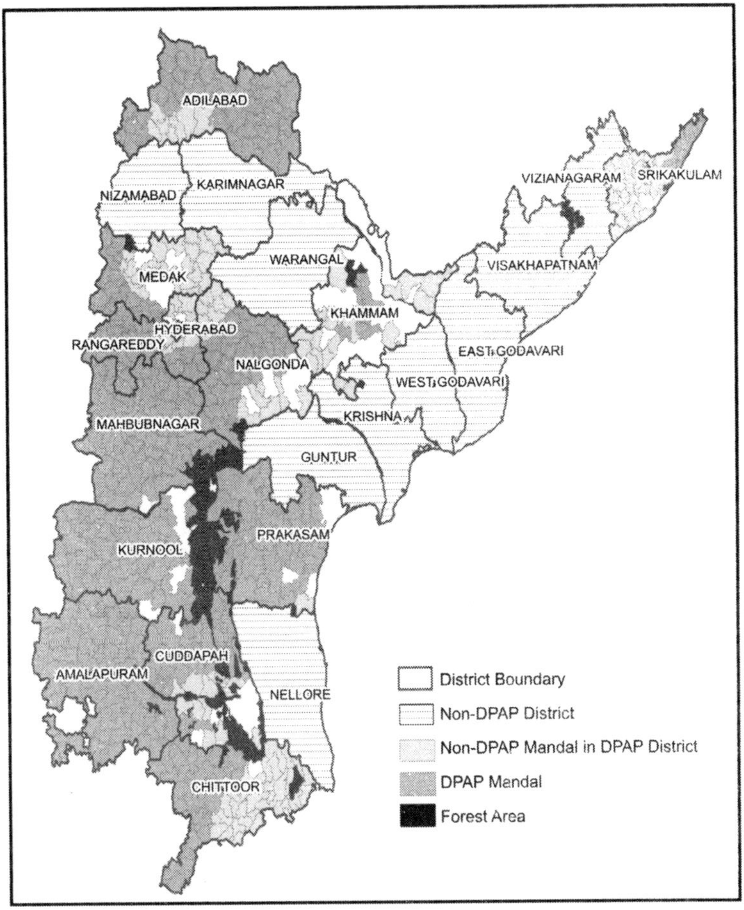

Source: GoAP, Commissioner of Rural Department (Watersheds), Hyderabad.

Map 19

Post-monsoon Temporal Water Level Zoning Maps

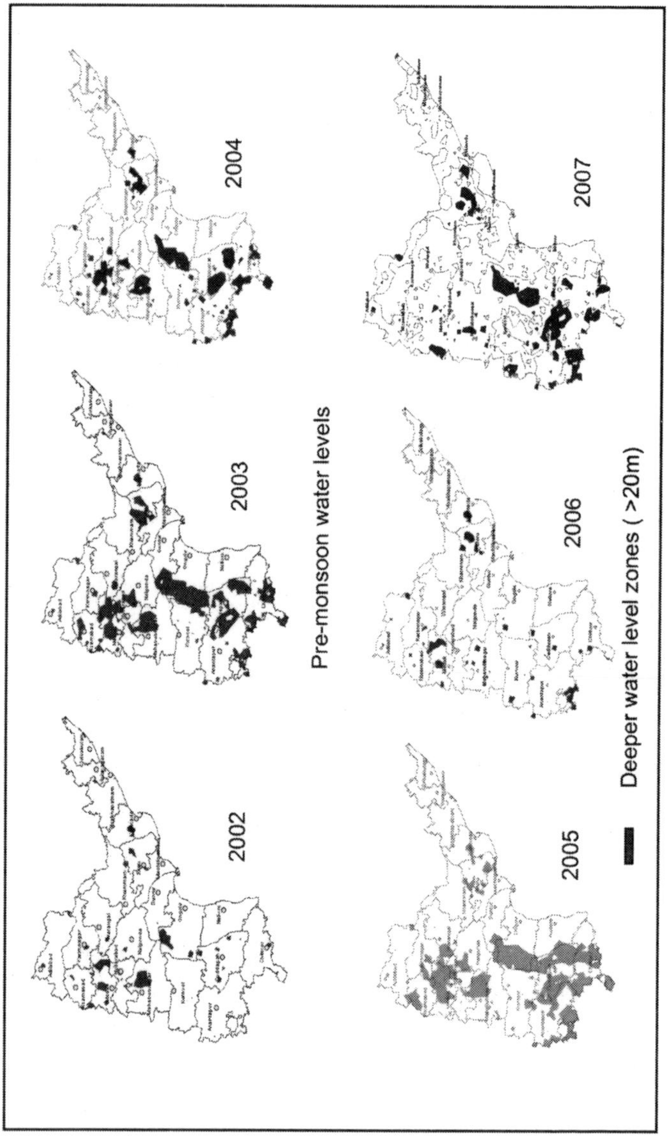

Map 20

Year-wise Post-monsoon Water Levels with Deep Water Level Zones

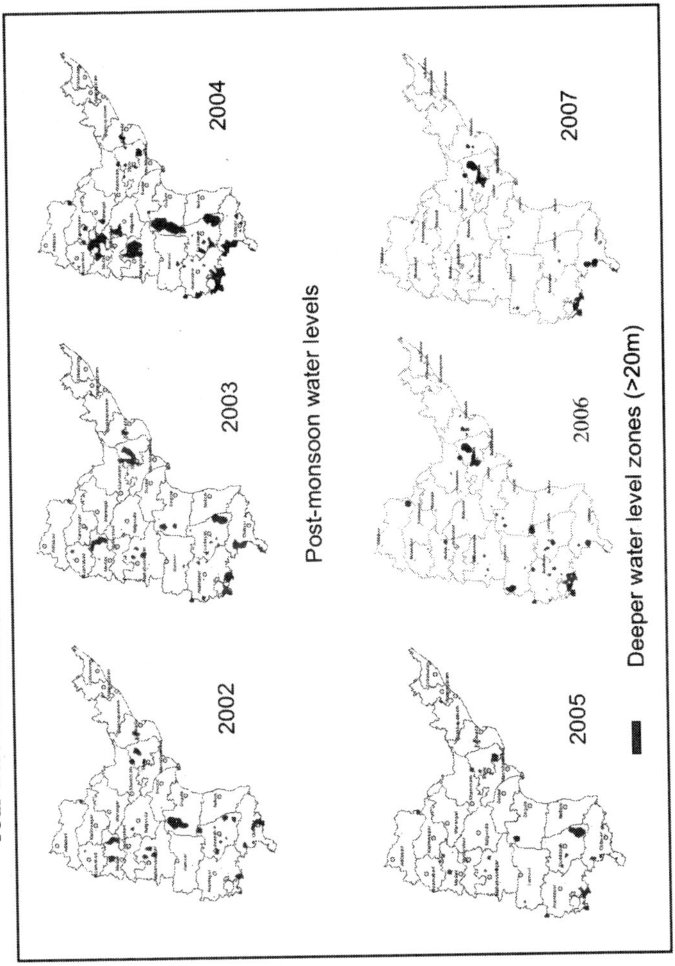

References

AFPRO (2006), 'A systematic Assessment of Community-based Groundwater Management Experiences in Andhra Pradesh', Action for Food Production, Field Unit 6, Hyderabad, December.

Alaerts, G., and Khoury, N. (2004), 'Arsenic Contamination of Groundwater: Mitigation Strategies and Policies', *Hydrogeol. J.*, 12 (1), pp. 103–115.

APFAMGS (2006), 'Rebuilding Groundwater Dependent Economy through Managed Aquifer Recharge (MAR)', Grama Vikas Samstha, Madanapally and APFAMGS, Hyderabad, January.

Bhatia, Bela (1992), 'Lush Fields and Parched Throats, Political Economy of Groundwater in Gujarat', *Economic and Political Weekly*, December 19–26, pp. A–142 to A–170.

Briscoe, J. (1999), 'Water Resources Management in Yemen: Results of a Consultation', Office Memorandum from John Briscoe, World Bank to Doris Köhn, Jean-Claude Villiard, Inder Sud, Salah Darghouth, Jamal Saghir and Mena Water Staff, Water Resources Sector Board Members, World Bank, Washington, US, November 1.

Burke, J., M. Moench, and C. Sauveplane (1999), 'Groundwater and Society: Problems in Variability and Points of Engagement'. In *Groundwater: Legal and Policy Perspectives*, (eds.), M. Salman and A. Salman, World Bank Technical Paper No. 456, pp. 31–52, Washington D.C.: The World Bank.

CGWB (2006), 'Dynamic Groundwater Resources of India' (as on March 2004), Central Ground Water Board, New Delhi.

CWC (1993), 'Estimates of Water Resources of India, Central Water Commission', Ministry of Water Resources, Government of India, New Delhi.

Deshpande, R.S. and V. Ratna Reddy (1990), 'Social Dynamics and Farmers' Society: A Case Study of *Pani Panchayat*' (jointly), *Indian Journal of Agricultural Economics,* Vol. XLV, No. 3, July–September.

Dhawan, B.D. (1995), 'Magnitude of Groundwater Exploitation', *Economic and Political Weekly*, Vol. No. 30, Issue No. 14.

FAO (2004), 'Economic Valuation of Water Resources in Agriculture: From the Sectoral to a Functional Perspective of Natural Resource Management', FAO Water Report 27, Rome.

FAO (2008), 'Evaluation of FAO Cooperation with India: Andhra Pradesh Farmer Managed Groundwater Systems (APFAMGS–GCP/IND/175/NET) Evaluation Report' (http://www.apfamgs.org), November.

Feitelson, E. (2006), 'Impediments to the Management of Shared Aquifers: A Political Economic Perspective', *Hydrogeology Journal* 14, pp. 319–329.

Fujita, K. and F. Hussain (1995), 'Role of the Groundwater Market in Agricultural Development and Income Distribution: A Case Study in a Northwest Bangladesh Village', *The Developing Economics*, XXXIII, pp. 442–463.

Gandhi, Vasant and Suresh Sharma (2009), 'The Performance of Rain-Water Harvesting Institutions in India: Empirical Study', In Lin Crase and Vasant P. Gandhi (ed.), *Reforming Institutions in Water Resource Management: Policy and Performance for Sustainable Development*, Earthscan: London.

GoAP (2003), 'AP Water Vision', Mission Support Unit of AP Water Conservation Mission, Andhra Pradesh Government.

GoAP (2004), 'AP Water Vision', Mission Support Unit of AP Water Conservation Mission, Andhra Pradesh Government.

GoAP (2006), 'Failed Bore Well Compensation Scheme', G.O. Rt No. 1099.

GoAP (2007), 'Participatory Groundwater in Andhra Pradesh Community Based Tank Management Project', Irrigation and CAD Department.

GoAP (2008), 'Groundwater Resource Andhra Pradesh 2007', State Ground Water Department, Hyderabad, August.

GoAP and GoI (2011), Dynamic Groundwater Resources of Andhra Pradesh (2008–09), State Government in Department, Hyderabad and Central Ground Water Board, Southern Region, Hyderabad, Vol. I and II, May.

GoI (1992), 'Model Bill to Regulate and Control the Development of Groundwater', New Delhi: Ministry of Water Resources.

GoI (1997), 'Groundwater Estimation Committee Groundwater Estimation Methodology 1997', Ministry of Water Resources, Government of India.

GoI (2002), 'Report on the Hard Rock Ground Water Estimation', Central Ground Water Board, Ministry of Water Resources, New Delhi.

GoI (2006), 'Central Ground Water Board', Dynamic Groundwater Resources of India (as on March 2004), New Delhi.

GoI (2007), 'Report of the Steering Committee on Water Resources for Eleventh Five Year Plan (2007–2012)', Planning Commission, May.

GoI (2007), 'Report of the Expert Group on Groundwater Management and Ownership', Planning Commission, New Delhi, September.

GoI (2009), 'Report of the Groundwater Resource Estimation Committee, Groundwater Resource Estimation Methodology', Ministry of Water Resources, Government of India, New Delhi.

GoI (2011), 'Report of the Working Group on Sustainable Groundwater Management as an Input to 12th Plan', Planning Commission, New Delhi, October 2.

GoI (2011), Dynamic Groundwater Resources of India, Central Ground Water Board, Faridabad, November.

IRM and ED (2008), 'Institutional Framework for Regulating Use of Ground Water in India', Final Report, Sponsored by Ministry of Water Resources Government of India, Delhi, September 16.

Janakarajan, S. (1993), 'Triadic Exchange Relations: An Illustration from South India', Institute of Development Studies, 24(3): 75–82.

Kemper, K.E. (2007), 'Instruments and Institutions for Groundwater Management,' In *The Agricultural Groundwater Revolution: Opportunities and Threats to Development*, edited by M. Giodano and K.G. Villholth, pp. 153–172, Oxford University Press.

Kumar, M. Dinesh (2005), 'Impact of Electricity Prices and Volumetric Water Allocation on Energy and Groundwater Demand Management: Analysis from Western India', *Energy Policy*, Vol. 33, pp. 39–51.

Kumar, M. Dinesh (2007), Groundwater Management in India: Physical, Institutional and Policy Alternatives, Sage Publications.

Kumar, M. Dinesh, Ankit Patel, R. Ravindranath, O.P. Singh (2008), 'Chasing a Mirage: Water Harvesting and Artificial Recharge in Naturally Water-Scarce Regions', *Economic and Political Weekly*, August 30.

Kumar, M. Dinesh, M.V.K. Sivamohan and A. Narayanamoorthy (2011), 'Turning India's Irrigation Portrait Upside Down: Dominant Views vs. Realities', *The Indian Economy Review*, Volume VII, January.

Kumar, M. Dinesh, M.V.K. Sivamohan, V. Niranjan and Nitin Bassi (2011), 'Groundwater Management in Andhra Pradesh: Time to Address Real Issues', Occasional Paper No. 4, Institute for Resource Analysis and Policy, Hyderabad.

Kumar, M. Dinesh, Christopher A. Scott and O.P. Singh (2011), Inducing the Shift from Flat-rate or Free Agricultural Power to Metered Supply: Implications for Groundwater Depletion and Power Sector Viability in India, *Journal of Hydrology*, Volume 409, Issues 1–2, 28 October 2011, pp. 382–394.

Lakhina, A.K. (2007), 'Sectoral Allocation of Groundwater and its Pricing', In *Groundwater Governance: Ownership of Groundwater and Pricing*, Saleem Romani et al. (eds),, Capital Publishing Company, New Delhi.

Llamas, and Martínez-Santos (2005), 'Intensive Groundwater Use: Silent Revolution and Potential Source of Social Conflicts', *Journal of Water Resources Planning and Management*, ASCE, September–October.

Malik, R.P.S. (nd), 'Energy Regulations as a Demand Management Option: Potentials, Problems and Prospects', Agro-economic Research Centre, Delhi University, Delhi.

Moench, Marcus (1992), 'Chasing the Water Table: Equity and Sustainability in Groundwater Management', *Economic and Political Weekly*, 27(51–52), pp. 171–177.

Mudarkartha, Srinivas (1999), *Status and Policy Framework of Groundwater in India*, VIKSAT.

Mukherji, A. (2007), 'The Energy-Irrigation Nexus and Its Implications for Groundwater Markets in Eastern Indo-Gangetic Basin: Evidence from West Bengal, India', *Energy Policy*, 35(12), pp. 6413–6430.

Pant, N. (2005), 'Control and Access to Groundwater in UP', *Economic and Political Weekly*, 40 (26), pp. 2672–2680, June 25–July 1.

Polak, P. (2004), 'Water and the Other Three Revolutions Needed to End Rural Poverty', Invited Paper in the World Water Week, Stockholm, 15–20 August, Preprint, Stockholm International Water Institute, Stockholm.

Reddy, M. Devender (2007), Water Management-Andhra Pradesh, Water Technology Centre, ANGRAU, Rajendranagar, Hyderabad.

Reddy, V. Ratna (2004), 'Managing Water Resources in India: A Synoptic Review', *Journal of Social and Economic Development*, Vol. 6, No. 2, July–December.

Reddy, V. Ratna (2005), 'Costs of Resource Depletion Externalities: A Study of Groundwater Overexploitation in Andhra Pradesh, India', *Environment and Development Economics* (Cambridge), Vol. 10, Part IV, August.

Reddy, V. Ratna (2012), 'Hydrological Externalities and Livelihoods Impacts: Informed Communities for Better Resource Management', *Journal of Hydrology*, Vols. 412–413, January 4.

Reddy, V. Ratna and Bhagirath Behra (2009), 'The Economic and Ecological Impacts of Tank Restoration in South India' (jointly), *European Journal of Development Research*, Vol. 21, February.

Reddy V. Ratna and Bhagirath Behera (2009), 'Economic Viability of Rehabilitation of Traditional Water Harvesting System: Empirical Evidence from South India', *Quarterly Journal of International Agriculture*, 48, Vol. 4, pp. 321–336

Reddy, V. Ratna and P. Prudhvikar Reddy (2005), 'How Participatory is Participatory Irrigation Management: A Study of Water User Associations in Andhra Pradesh', *Economic and Political Weekly*, Vol. XL, No. 53, December 31.

Reddy, V. Ratna, M. Gopinath Reddy and John Soussan (2010), *Political Economy of Watershed Management: Politics, Institutions, Implementation and Livelihoods*, Rawat Publications, Jaipur.

Reddy, V. Ratna and S. Galab (2006), 'Agrarian Crisis: Looking Beyond Debt Trap', *Economic and Political Weekly*, Vol. 41, No. 19, pp. 1838–40, May 13–19.

Reddy, V. Ratna, Geoff Syme, Ram Ranjan, Paul Pavelic, M. Srinivasa Reddy, Sanjit Kumar Rout and Acharya Sreedhar (2011), 'Scale Issues in Meso-Watershed Development: Farmers' Perceptions on Designing and Implementing the Common Guidelines', Working

Paper No. 2, Livelihoods and Natural Resource Management Institute, Hyderabad.

Roy, K.C. (1989), Optimization of Unconfined Shallow Aquifer Water Storage for Irrigation, Unpublished PhD dissertation, Utah State University.

Rural Integrated Development Society (2011), Social Regulation Project of Madirepalli, Power Point Presentation, RIDS, Garladinne Mandal, Anantapur District.

Saleth, R.M. (1994), 'Groundwater Markets in India: A Legal and Institutional Perspective,' *Indian Economic Review*, 29(2), pp. 157–176.

Saleth, R.M. (2005), 'Water Rights and Entitlements in India', Instrument Paper prepared for the World Bank-India Office for developing World Bank's Country Water Resources Sector Strategy for India, Colombo, Sri Lanka: International Water Management Institute, p. 40, February 8.

Sandoval, R. (2004), 'A Participatory Approach to Integrated Aquifer Management: The Case of Guanajuato State, Mexico', In Kemper, K.E. (ed.) *Groundwater: From Development to Management*, Hydrogeology Journal Theme Issue, Springer, New York: The World Bank, Washington, DC.

Scott, C.A. and T. Shah (2004), 'Groundwater Overdraft Reduction Through Agricultural Energy Policy: Insights from India and Mexico', *International Journal of Water Resources Development*, 20(2), pp. 149–164.

Shah, T. (1989), 'Groundwater Grids in the Villages of Gujarat: Evolution, Structure, Working and Impacts', *Wamana*, January, pp. 14–29.

Shah, T. (1993), *Groundwater Markets and Irrigation Development, Bombay*, Oxford University Press.

Shah, T. (2004), 'Groundwater and Human Development: Challenges and Opportunities in Livelihoods and Environment', Invited Paper in the World Water Week, Stockholm, 15–20 August, Preprint, Stockholm International Water Institute, Stockholm.

Shah, T. (2007), 'Crop per Drop of Diesel? Energy Squeeze on India's Smallholder Irrigation', *Economic and Political Weekly*, September 29.

Shah, T. (2009), *Taming the Anarchy: Groundwater Governance in South Asia* (Washington DC: The Resources for the Future Press).

Shah, T., Giordano, M. and Wang, J. (2004a), 'Irrigation Institutions in a Dynamic Economy: What is China Doing Differently from India?', *Economic and Political Weekly* XXXIX (31), pp. 3452–3461.

Shah, T., Hussain, I., Saeed-ur-Rehman (2000), *Irrigation Management in Pakistan and India: Comparing Notes on Institutions and Policies,* Colombo, Sri Lanka: International Water Management Institute (IWMI), IWMI Working Paper 4, p. 21.

Shah, T., Koppen, B.V., Merry, D., Samad, M. (2002), 'Institutional Alternatives in African Smallholder Irrigation: Lessons from International Experience with Irrigation Management Transfer', Colombo, Srilanka: IWMI. V, p. 24 (IWMI Research Report 60).

Shankar, K. (1992), *Dynamics of Groundwater Irrigation*, New Delhi, Segment Books.

Sharma, K.D. (2009), Groundwater Management for Food Security, *Current Science*, Vol. 96, No. 11, pp. 1444–1447.

Sharma, S.C. (1995), 'Regulation of Groundwater Development in India: Existing Provisions and Future Options', In *Groundwater Law: The Growing Debate*, M. Moench (ed.), VIKSAT Pacific Institute Collaborative Groundwater Project, Ahmedabad.

Singh, Katar (1995), 'Co-operative Property Rights as an Instrument of Managing Groundwater', In *Groundwater Law: The Growing Debate*, M. Moench (ed.), VIKSAT Pacific Institute Collaborative Groundwater Project, Ahmedabad.

Swallow, B., Garritty, D.P. and van Noordwijk, M. (2001), 'The Effect of Scales, Flows and Filters on Property Rights and Collective Action in Watershed Management', Capri Working Paper No. 16, EFPRI, Washington, D.C.

Syme, J. Geoffrey, V. Ratna Reddy, Paul Pavelic, Barry Croke, Ram Ranjan (2011), 'Confronting Scale in Watershed Development in India', *Hydrogeology Journal*, 17 January.

The World Bank (2000), Middle East and North Africa: Regional Water Initiative Workshop on Sustainable Groundwater Management, Sana, Yemen, 25–28 June.

The World Bank (2010), Deep Wells and Prudence: Owards Pragmatic Action for Addressing Groundwater Overexploitation in India, Washington, D.C.

Valencia Statement (2004), *Selected Papers of the International Symposium on Intensive Use of Groundwater: Challenges and*

Opportunities, Sahuquillo et al., eds., Balkema, Dordrecht, The Netherlands, pp. 399–400.

Van Halsema, G.E. (2002), 'Trial and Re-trial: The Evolution of Irrigation Modernisation in NWFP, Pakistan', PhD diss., Wageningen University, Department of Water Resources, Netherlands.

Villarroya, F. and Aldwell, C.R. (1998), 'Sustainable Development and Groundwater Resources Exploitation', *Environ. Geol.* 34 (2–3), pp. 111–115.

Wang, J., Huang, J., Blanke, A., Huang, Q. and Rozelle, S. (2007), 'The Development, Challenges and Management of Groundwater in Rural China', In Mark Giordano, Villholth, K.G. (eds.), *The Agricultural Groundwater Revolution: Comprehensive Assessment of Water Management in Agriculture*, Wallingford, Oxfordshire, UK, pp. 37–62.

World Bank (nd), 'Deep Wells and Prudence: Towards Pragmatic Action for Addressing Groundwater Overexploitation in India', APFAMGS, http://www.apfamgs.org.

Index